网络数据采集技术
Java网络爬虫实战

钱 洋 姜元春 著

电子工业出版社
Publishing House of Electronics Industry
北京·BEIJING

内 容 简 介

本书以 Java 为开发语言，系统地介绍了网络爬虫的理论知识和基础工具，包括网络爬虫涉及的 Java 基础知识、HTTP 协议基础与网络抓包、网页内容获取、网页内容解析和网络爬虫数据存储等。本书选取典型网站，采用案例讲解的方式介绍网络爬虫中涉及的问题，以增强读者的动手实践能力。同时，本书还介绍了 3 种 Java 网络爬虫开源框架，即 Crawler4j、WebCollector 和 WebMagic。

本书适用于 Java 网络爬虫开发的初学者和进阶者；也可作为网络爬虫课程教学的参考书，供高等院校文本挖掘、自然语言处理、大数据商务分析等相关学科的本科生和研究生参考使用；也可供企业网络爬虫开发人员参考使用。

未经许可，不得以任何方式复制或抄袭本书之部分或全部内容。
版权所有，侵权必究。

图书在版编目（CIP）数据

网络数据采集技术：Java 网络爬虫实战 / 钱洋，姜元春著. —北京：电子工业出版社，2020.1
ISBN 978-7-121-37607-8

Ⅰ. ①网... Ⅱ. ①钱... ②姜... Ⅲ. ①JAVA 语言－程序设计 Ⅳ. ①TP312.8

中国版本图书馆 CIP 数据核字（2019）第 219551 号

责任编辑：林瑞和　　　特约编辑：田学清
印　　刷：天津千鹤文化传播有限公司
装　　订：天津千鹤文化传播有限公司
出版发行：电子工业出版社
　　　　　北京市海淀区万寿路 173 信箱　　邮编：100036
开　　本：720×1000　1/16　印张：23.75　字数：478.8 千字
版　　次：2020 年 1 月第 1 版
印　　次：2020 年 1 月第 1 次印刷
定　　价：79.00 元

凡所购买电子工业出版社图书有缺损问题，请向购买书店调换。若书店售缺，请与本社发行部联系，联系及邮购电话：(010) 88254888，88258888。

质量投诉请发邮件至 zlts@phei.com.cn，盗版侵权举报请发邮件到 dbqq@phei.com.cn。
本书咨询联系方式：010-51260888-819，faq@phei.com.cn。

前　言

近几年，网络空间大数据（Big Data）已成为各领域研究的热门话题。在企业应用方面，天猫利用海量的用户数据挖掘年轻消费者偏好，并将用户偏好反馈给手机研发部门，将其用于手机设计；汽车之家利用平台中用户生成的大数据对用户进行画像，在此基础上开展个性化营销。在学术界，很多领域的学者针对大数据衍生出的新问题开展学术研究，如大数据驱动的客户洞察、大数据驱动的个性化推荐、大数据驱动的管理决策等。

在网络大数据环境下，数据采集尤为重要。因此，很多企业都提供了（高级）数据采集工程师的职位。对于很多在校大学生而言，尤其是硕士生和博士生，网络数据采集是一项必备的技能。

在编写本书之前，笔者主要从事数据采集系统的设计与开发工作。在 CSDN 社区上，笔者撰写过一系列介绍 Java 网络爬虫的博客，这些博客为笔者的主页带来了不少访问量；同时，也有许多博客读者通过邮件的方式，向笔者咨询网络爬虫相关的工具使用、程序调试等问题。为此，笔者对 Java 网络爬虫所涉及的知识与技术进行了系统的梳理，并打算编写一本关于 Java 网络爬虫的书籍。在写作过程中，笔者与具有丰富网络爬虫教学经验的姜元春教授就写作逻辑、介绍的知识点、使用的案例等多方面的内容进行了多次讨论。本书的内容更加注重爬虫理论、开发基础与实战演练。基于对本书爬虫案例的研读，读者可以快速开发自己需要的其他网络爬虫程序。

本书的内容

本书分为 9 章，具体内容如下所示。

第 1 章至第 3 章：这 3 章重点介绍与网络爬虫开发相关的基础知识，其中包括网络爬虫的原理、Java 基础知识和 HTTP 协议等内容。

第 4 章至第 6 章：这 3 章分别从网页内容获取、网页内容解析和网络爬虫数据存储 3 个方面介绍网络爬虫开发过程中所涉及的一系列技术。在这 3 章中，涉及很多开源工具的使用，如 Jsoup、HttpClient、HtmlCleaner、Fastjson、POI3 等。

第 7 章：本章利用具体的实战案例，讲解网络爬虫开发的流程。通过对本章的学习，读者可以轻松开发 Java 网络爬虫。

第 8 章：针对一些复杂的页面，如动态加载的页面（执行 JavaScript 脚本），本章介绍了一款实用的工具——Selenium WebDriver。

第 9 章：本章重点介绍了 3 种比较流行的 Java 网络爬虫开源框架，即 Crawler4j、WebCollector 和 WebMagic。读者可根据数据采集需求，自行开发支持多线程采集、断点采集、代理切换等功能的网络爬虫项目。

本书的特色

- 注重基础：俗话说，基础不牢，地动山摇。本书从可读性和实用性出发，重点介绍了网络爬虫中涉及的基础知识。
- 系统性：本书系统地梳理了网络爬虫的逻辑和开发网络爬虫需要掌握的技术。对网络爬虫初学者和进阶者而言，学习这些内容将有利于解决数据采集过程中遇到的各种问题。
- 详细的案例讲解：本书选取了较为典型的网站，讲解网络爬虫经常遇到的问题，如 HTTPS 请求认证问题、大文件内容获取问题、模拟登录问题、不同格式文件（文本、图片和 PDF 等）的存储问题、定时数据采集问题等。
- 开源框架：本书介绍了 3 种 Java 网络爬虫开源框架，即 Crawler4j、WebCollector 和 WebMagic。通过对这 3 种网络爬虫开源框架的学习，读者可以轻松开发一些高性能的网络爬虫项目。
- 完整的代码：为便于读者学习，对于每个数据网络爬虫项目，笔者都提供了完整的代码，并且在代码中给出了清晰的注释。

适合的读者

- Java 网络爬虫开发的初学者和进阶者。
- 科研人员，尤其是从事网络大数据驱动研究的硕士生和博士生。
- 开设相关课程的高等院校的师生。
- 企业网络爬虫开发人员。

说明

网络爬虫作为一项技术，更应该服务于社会。在使用该技术的过程中，应遵守 Robots 协议（互联网行业数据抓取的道德协议）。同时，需要注意对数据所涉及的知识产权和隐私信息进行保护。另外，采集数据时，需要注意礼貌，即不频繁地请求网页，以防止给数据提供者的服务器造成不良影响。在使用所采集的数据时，需要注意是否涉及商业利益和相关法律。最后，本书中所有使用的案例皆为测试案例，仅供读者学习使用，本书中的 URL 均做了处理。

基金项目

本书由国家自然科学基金重大项目课题"面向大数据的商务分析与计算方法以及支撑平台研究（71490725）"、国家自然科学基金重大研究计划子课题"面向商务领域的大数据资源池及集成示范平台（91746302）"、国家自然科学基金优秀青年基金"个性化营销理论与方法（71722010）"提供资助。

勘误

由于笔者的水平有限，书中难免出现一些错误及不准确之处，恳请读者批评指正。为及时更正书中不恰当的内容，笔者在 CSDN 博客中创建了一个板块，读者可以将书中的问题以评论的方式进行反馈，笔者将针对这些问题进行勘误。另外，也欢迎读者通过发送电子邮件（qy20115549@126.com）的方式，反馈书稿的问题。

致谢

感谢电子工业出版社的林瑞和编辑、合肥工业大学电子商务研究所的刘业政教授和孙见山副教授等给本书提出的宝贵建议。

感谢华为的杜非、王佳佳和王锦坤师兄的帮助，是他们将我带入编程的世界。

感谢淮南师范学院的孙娜丽女士对整本书稿写作语言的梳理。

感谢合肥工业大学电子商务研究所的朱婷婷、杨露、田志强、宋颖欣、张雪、李哲、贺菲菲、叶畅、陶守正、梁瑞诚等博士参与本书内容的讨论。

最后，希望热爱网络爬虫开发的小伙伴们能够喜欢本书。

钱 洋

2019 年 9 月

【读者服务】

微信扫码回复：（37607）

- 获取本书配套*代码*资源
- 获取更多技术专家分享视频与学习资源
- 加入读者交流群，与更多读者互动、与本书作者互动

目 录

第 1 章 网络爬虫概述与原理 1
    1.1 网络爬虫简介 1
    1.2 网络爬虫分类 2
    1.3 网络爬虫流程 4
    1.4 网络爬虫的采集策略 5
    1.5 学习网络爬虫的建议 5
    1.6 本章小结 6

第 2 章 网络爬虫涉及的 Java 基础知识 7
    2.1 开发环境的搭建 7
        2.1.1 JDK 的安装及环境变量配置 7
        2.1.2 Eclipse 的下载 9
    2.2 基本数据类型 10
    2.3 数组 11
    2.4 条件判断与循环 12
    2.5 集合 15
        2.5.1 List 和 Set 集合 15
        2.5.2 Map 集合 16
        2.5.3 Queue 集合 17
    2.6 对象与类 19
    2.7 String 类 21
    2.8 日期和时间处理 23
    2.9 正则表达式 26
    2.10 Maven 工程的创建 29
    2.11 log4j 的使用 33
    2.12 本章小结 40

第3章　HTTP协议基础与网络抓包 41

3.1　HTTP协议简介 41
3.2　URL 42
3.3　报文 44
3.4　HTTP请求方法 46
3.5　HTTP状态码 46
3.5.1　状态码2XX 47
3.5.2　状态码3XX 47
3.5.3　状态码4XX 48
3.5.4　状态码5XX 48
3.6　HTTP信息头 48
3.6.1　通用头 49
3.6.2　请求头 52
3.6.3　响应头 55
3.6.4　实体头 56
3.7　HTTP响应正文 57
3.7.1　HTML 58
3.7.2　XML 60
3.7.3　JSON 61
3.8　网络抓包 64
3.8.1　简介 64
3.8.2　使用情境 65
3.8.3　浏览器实现网络抓包 65
3.8.4　其他网络抓包工具推荐 70
3.9　本章小结 70

第4章　网页内容获取 71

4.1　Jsoup的使用 71
4.1.1　jar包的下载 71
4.1.2　请求URL 72
4.1.3　设置头信息 75
4.1.4　提交请求参数 78
4.1.5　超时设置 80

4.1.6　代理服务器的使用 81
　　　4.1.7　响应转输出流（图片、PDF 等的下载）.. 83
　　　4.1.8　HTTPS 请求认证 85
　　　4.1.9　大文件内容获取问题 89
　4.2　HttpClient 的使用 91
　　　4.2.1　jar 包的下载 91
　　　4.2.2　请求 URL 92
　　　4.2.3　EntityUtils 类 97
　　　4.2.4　设置头信息 98
　　　4.2.5　POST 提交表单 100
　　　4.2.6　超时设置 103
　　　4.2.7　代理服务器的使用 105
　　　4.2.8　文件下载 106
　　　4.2.9　HTTPS 请求认证 108
　　　4.2.10　请求重试 111
　　　4.2.11　多线程执行请求 114
　4.3　URLConnection 与 HttpURLConnection 117
　　　4.3.1　实例化 117
　　　4.3.2　获取网页内容 118
　　　4.3.3　GET 请求 118
　　　4.3.4　模拟提交表单（POST 请求）.. 119
　　　4.3.5　设置头信息 120
　　　4.3.6　连接超时设置 121
　　　4.3.7　代理服务器的使用 122
　　　4.3.8　HTTPS 请求认证 122
　4.4　本章小结 124

第 5 章　网页内容解析 125
　5.1　HTML 解析 125
　　　5.1.1　CSS 选择器 125
　　　5.1.2　Xpath 语法 127
　　　5.1.3　Jsoup 解析 HTML 128
　　　5.1.4　HtmlCleaner 解析 HTML 135

5.1.5 HTMLParser 解析 HTML 139
5.2 XML 解析 144
5.3 JSON 解析 145
5.3.1 JSON 校正 145
5.3.2 org.json 解析 JSON 147
5.3.3 Gson 解析 JSON 152
5.3.4 Fastjson 解析 JSON 157
5.3.5 网络爬虫实战演练 159
5.4 本章小结 165

第 6 章 网络爬虫数据存储 166
6.1 输入流与输出流 166
6.1.1 简介 166
6.1.2 File 类 166
6.1.3 文件字节流 169
6.1.4 文件字符流 172
6.1.5 缓冲流 176
6.1.6 网络爬虫下载图片实战 180
6.1.7 网络爬虫文本存储实战 184
6.2 Excel 存储 188
6.2.1 Jxl 的使用 188
6.2.2 POI 的使用 191
6.2.3 爬虫案例 198
6.3 MySQL 数据存储 202
6.3.1 数据库的基本概念 203
6.3.2 SQL 语句基础 203
6.3.3 Java 操作数据库 207
6.3.4 爬虫案例 217
6.4 本章小结 219

第 7 章 网络爬虫实战项目 220
7.1 新闻数据采集 220
7.1.1 采集的网页 220
7.1.2 框架介绍 222

		7.1.3　程序编写 223
	7.2　企业信息采集 235
		7.2.1　采集的网页 235
		7.2.2　框架介绍 238
		7.2.3　第一层信息采集 239
		7.2.4　第二层信息采集 248
	7.3　股票信息采集 256
		7.3.1　采集的网页 256
		7.3.2　框架介绍 257
		7.3.3　程序设计 258
		7.3.4　Quartz 实现定时调度任务 267
	7.4　本章小结 271

第 8 章　Selenium 的使用 272
	8.1　Selenium 简介 272
	8.2　Java Selenium 环境搭建 272
	8.3　浏览器的操控 274
	8.4　元素定位 276
		8.4.1　id 定位 276
		8.4.2　name 定位 277
		8.4.3　class 定位 278
		8.4.4　tag name 定位 278
		8.4.5　link text 定位 278
		8.4.6　Xpath 定位 279
		8.4.7　CSS 选择器定位 279
	8.5　模拟登录 280
	8.6　动态加载 JavaScript 数据（操作滚动条） 283
	8.7　隐藏浏览器 285
	8.8　截取验证码 287
	8.9　本章小结 291

第 9 章　网络爬虫开源框架 292
	9.1　Crawler4j 的使用 292
		9.1.1　Crawler4j 简介 292

- 9.1.2 jar 包的下载 292
- 9.1.3 入门案例 293
- 9.1.4 相关配置 297
- 9.1.5 图片的采集 300
- 9.1.6 数据采集入库 304
- 9.2 WebCollector 的使用 312
  - 9.2.1 WebCollector 简介 312
  - 9.2.2 jar 包的下载 313
  - 9.2.3 入门案例 313
  - 9.2.4 相关配置 318
  - 9.2.5 HTTP 请求扩展 319
  - 9.2.6 翻页数据采集 327
  - 9.2.7 图片的采集 331
  - 9.2.8 数据采集入库 334
- 9.3 WebMagic 的使用 347
  - 9.3.1 WebMagic 简介 347
  - 9.3.2 jar 包的下载 347
  - 9.3.3 入门案例(翻页数据采集) 347
  - 9.3.4 相关配置 351
  - 9.3.5 数据存储方式 352
  - 9.3.6 数据采集入库 355
  - 9.3.7 图片的采集 365
- 9.4 本章小结 368

第 1 章

网络爬虫概述与原理

1.1 网络爬虫简介

随着互联网的迅速发展，网络数据资源呈爆炸式增长，信息需求者如何从网络中提取信息变得更加重要。如今，有效地获取网络数据资源的方式，便是网络爬虫。网络爬虫（Web Crawler）又称为网络蜘蛛（Web Spider）或 Web 信息采集器，是一种按照指定规则，自动抓取或下载网络资源的计算机程序或自动化脚本。

对网络爬虫狭义上的理解：利用标准网络协议（如 HTTP、HTTPS 等），根据网络超链接和信息检索方法（如深度优先）遍历网络数据的软件程序。

对网络爬虫功能上的理解：确定待采集的 URL 队列，获取每个 URL 对应的网页内容（如 HTML 和 JSON 等），根据用户要求解析网页中的字段（如标题），并存储解析得到的数据。

网络爬虫技术在搜索引擎中扮演着信息采集器的角色，是搜索引擎模块中的最基础的部分。例如，我们常用的搜索引擎 Google、百度、必应（Bing）都采用网页爬虫技术采集海量的互联网数据。图 1.1 展示了搜索引擎的大致结构。第一步，利用网络爬虫技术自动化地采集互联网中的网页信息。第二步，存储采集的信息。在存储过程中，往往需要检测重复内容，从而避免大量重复信息的采集；同时，网页之间的链接关系也需要存储，原因是链接关系可用来计算网页内容的重要性。第三步，数据预处理操作，即提取文字、分词、消除噪音以及链接关系计算等。第四步，对预处理的数据建立索引库，方便用户快速查找，常用的索引方法有后缀数组、签名文件和倒排文件。第五步，基于用户检索的内容（如用户输入的关键词），搜索引擎从网页索引库中查找符合该关键词的所有网页（结果集），通过对结果集的排序，将最相关的网页返回给用户。

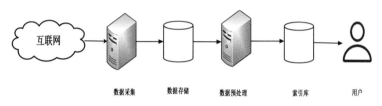

图 1.1 搜索引擎大致结构

另外，网络爬虫在其他方面也发挥着重要作用。

大数据环境下舆情分析与监测：政府或企业基于网络爬虫技术，采集论坛评论、在线博客、新闻媒体和微博等网站中的海量数据，采用数据挖掘相关方法（如实体识别、词频统计、文本情感计算、主题识别与演化等），发掘舆情热点、跟踪目标话题，并根据一定的标准采取相应的舆情控制与引导措施。

大数据环境下的用户分析：企业利用网络爬虫技术，采集用户基本信息、用户对企业或商品的看法、观点以及态度等数据、用户之间的互动信息等。基于这些信息，企业可以对用户进行画像，如用户基本属性画像、用户产品特征画像、用户互动特征画像等，发掘用户对产品的个性化偏好与需求。同样，也可分析企业自身产品的优势和顾客反馈情况等。

科研需求：针对网络大数据驱动、多源异构数据驱动的科学研究，必然涉及网络数据采集技术。例如，针对网络中的多源异构数据（如数字、文本、图片和视频等），如何更好地管理与存储所采集的数据、如何进行数据的过滤与融合、如何对数据的可用性进行评估、如何将数据应用到商业分析中等，都是目前研究的热点问题。

1.2 网络爬虫分类

网络爬虫按照系统结构和实现技术，大致可分为 4 类，即通用网络爬虫、聚焦网络爬虫、增量网络爬虫和深层网络爬虫。

通用网络爬虫：又称为全网网络爬虫，其在采集数据时，由部分种子 URL 扩展到整个网络的全部页面，主要应用于搜索引擎数据的采集。这类网络爬虫的数据采集范围较广，数据采集量巨大，对数据采集的速度和存储空间有较高的要求，通常需要深度遍历网站的资源。例如，Apache 的子项目 Nutch 便是一个高效的通用网络爬虫框架，其使用分布式的方式采集数据。有兴趣详细学习 Nutch 框架使用的读者，可参考书籍 *Web Crawling and Data Mining with Apache Nutch*。

聚焦网络爬虫：又称为主题网络爬虫，是指选择性地采集那些与预先定义好的主题相关的页面。相比于通用网络爬虫，聚焦网络爬虫采集的网页资源少，主要用于满足特定人群对特定领域信息的需求。在聚焦网络爬虫中，需要设计过滤策略，即过滤与所定主题无关的页面。我们可以将用户查询的关键词作为主题，包含该关键词的页面视为主题相关，如图 1.2 所示。

图 1.2 采集关键词相关网页

增量网络爬虫：是指对已下载网页采取增量式更新，只采集新产生的或者已经发生变化网页的爬虫。增量网络爬虫能够在一定程度上保证所爬行的页面是尽可能新的页面，历史已经采集过的页面不重复采集。增量网络爬虫避免了重复采集数据，可以减小时间和空间上的耗费。针对小规模特定网站的数据采集，在设计网络爬虫时，可构建一个基于时间戳判断是否更新的数据库，通过判断时间戳的先后，判断程序是否继续采集，同时更新数据库中的时间戳信息。图 1.3 为某论坛的帖子的页面，如果需要每隔一段时间采集最新发布的帖子，便可以通过对比当前时间（如 2018 年 9 月 25 日）和数据库中封存的上次数据采集的时间（如 2018 年 8 月 25 日），进而确定采集到论坛第几页便终止后面页数的数据采集。

图 1.3 增量网络爬虫案例

深度网络爬虫：即 Deep Web 爬虫，指对大部分内容不能通过静态链接获取，只有用户提交表单信息才能获取 Web 页面的爬虫。

1.3 网络爬虫流程

普通网络爬虫的流程大致如图 1.4 所示，一般包含 URL 队列模块、页面内容获取模块、页面解析模块、数据存储模块和 URL 过滤模块。具体流程可描述如下。

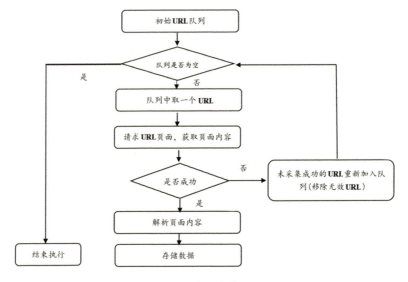

图 1.4　网络爬虫的流程

步骤 1，选取部分种子 URL（或初始 URL），将其放入待采集的队列中。如在 Java 中，可以放入 List、LinkedList 及 Queue 中。

步骤 2，判断 URL 队列是否为空，如果为空则结束程序的执行，否则执行步骤 3。

步骤 3，从待采集的 URL 队列中取出一个 URL，获取 URL 对应的网页内容。在此步骤需要使用 HTTP 响应状态码（如 200 和 403 等）判断是否成功获得了数据，如响应成功则执行解析操作；如响应不成功，则将其重新放入待采集 URL 队列（注意这里需要过滤掉无效 URL）。

步骤 4，针对响应成功后获取的数据，执行页面解析操作。此步骤根据用户需求获取网页内容中的部分字段，如汽车论坛帖子的 id、标题和发表时间等。

步骤 5，对步骤 4 解析的数据执行数据存储操作。

1.4 网络爬虫的采集策略

网络爬虫的采集策略一般分为两种：深度优先搜索（Depth-First Search）策略和广度优先搜索（Breadth-First Search）策略。

深度优先搜索策略：从根节点开始，根据优先级向下遍历该根节点对应的子节点。当访问到某一子节点时，以该子节点为入口，继续向下层遍历，直到没有新的子节点可以继续访问为止。接着使用回溯法，找到没有被访问到的节点，以类似的方式进行搜索。图 1.5 给出了理解深度优先搜索的一个简单案例。

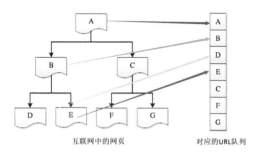

图 1.5　深度优先搜索遍历案例

广度优先搜索策略：又称为宽度优先搜索策略，从根节点开始，沿着网络的宽度遍历每一层的节点，如果所有节点均被访问，则终止程序。图 1.6 给出了理解广度优先搜索的一个简单案例。基于广度优先的爬虫是最简单的采集网站信息的采集器，也是目前使用较为广泛的采集器。

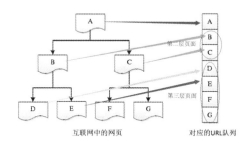

图 1.6　广度优先搜索遍历案例

1.5 学习网络爬虫的建议

很多学习 Java 编程知识的学生，采取的学习方式往往是看书或者观看视频，而这将导致面对具体的实战项目时，不知道如何上手。另外，一些学生学习了书本或视频中的 Java 编程知识，却不知道该怎么应用这些知识。为此，笔者建议读者通过网络爬虫项目的编写，加深对 Java 编程知识的使用。对于零基础入门 Java 网络爬虫的读者，在学习中需要注意以下几点。

Java 基础知识的掌握：了解 Java 方面的基础知识，如基本数据类型、运算符、判断语句、循环语句、数组和集合操作等。这部分内容一般通过大学教材便可学习和掌握。

网络爬虫原理理解：理解了网络爬虫原理，便可以发现学习网络爬虫会涉及网页数据的请求、网页数据的解析以及网页数据的存储等一系列操作。本书的写作顺序便是依据网络爬虫的原理和操作流程进行的。

吃透基本爬虫代码：在 Java 网络爬虫中，涉及许多开源 jar 包的使用，如网页请求工具 HttpClient、JSON 解析工具 Fastjson、数据库操作工具 QueryRunner 等。熟练掌握这些工具的使用，将方便我们快速开发网络爬虫。另外，在本书中介绍了大量网络爬虫实战案例，读者可以从 GitHub 网站上将代码下载到本地，进行实际演练，并根据自身需求，改写程序。

学会搜索：在实战开发以及程序调试的过程中，读者可能会遇到许多问题。解决这些问题最好的方式，便是网络搜索。例如，使用 HttpClient 直接请求某一个 URL 抛出"PKIX path building failed"的错误，我们便可以通过 Google 或者百度搜索该错误的原因以及解决方案。

1.6 本章小结

首先，本章介绍了网络爬虫基本概念及应用。其次，介绍了网络爬虫的分类（通用网络爬虫、聚焦网络爬虫、增量网络爬虫和深层网络爬虫）。接着，对网络爬虫的流程做了详细的讲解，并介绍了网络爬虫的两种采集策略（深度优先和宽度优先）。最后，笔者给出了学习网络爬虫的一些建议。

第 2 章

网络爬虫涉及的 Java 基础知识

2.1 开发环境的搭建

2.1.1 JDK 的安装及环境变量配置

学习和使用 Java，需要下载和安装 Java 开发工具包 JDK。

本书安装和使用的 JDK 版本为 JDK 8，操作系统为 Windows10 64 位。

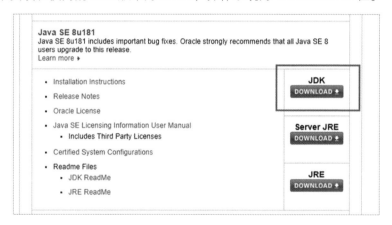

图 2.1 JDK 下载页面

单击如图 2.1 所示页面的下载按钮（DOWNLOAD），跳转到如图 2.2 所示的下载页面，选择接受许可。同时，根据读者使用的操作系统，选择下载相应的 JDK，笔者下载的 JDK 版本为 jdk-8u181-windows-x64.exe。

JDK 下载完成后，双击 jdk-8u181-windows-x64.exe，根据提示信息进行安装。安装过程中，读者可自行选择安装路径，如笔者选择安装的目录为 G:\software\Java。在安装 JDK 的同时，也安装了 JRE，如图 2.3 所示。

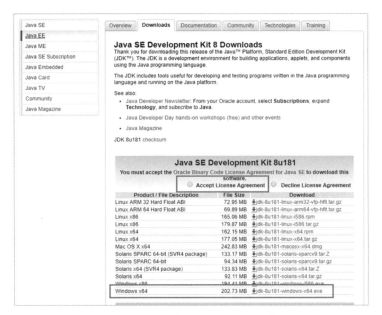

图 2.2　不同系统对应的 JDK 下载

图 2.3　选择安装目录

JDK 提供了 Java 编译器（javac.exe）和 Java 解释器（java.exe），其位于 Java 安装目录的\bin 文件目录中，为了更加方便地使用编译器以及解释器，需要在系统环境变量中添加相应的目录。在 Windows 10 中，依次单击"此电脑"→"高级系统设置"，在"高级"选项卡中单击"环境变量"，出现如图 2.4 所示界面。在系统变量中，添加如下变量名并设置变量值。

```
JAVA_HOME: G:\software\Java\jdk1.8.0_65
CLASSPATH: .;%JAVA_HOME%\lib\dt.jar;%JAVA_HOME%\lib\tools.jar;
Path: G:\software\Java\jdk1.8.0_65\bin
```

在 cmd（Command，命令提示符）中，输入 java -version 命令，如果出现如图 2.5 所示的内容，说明环境变量配置成功。

图 2.4　配置环境变量

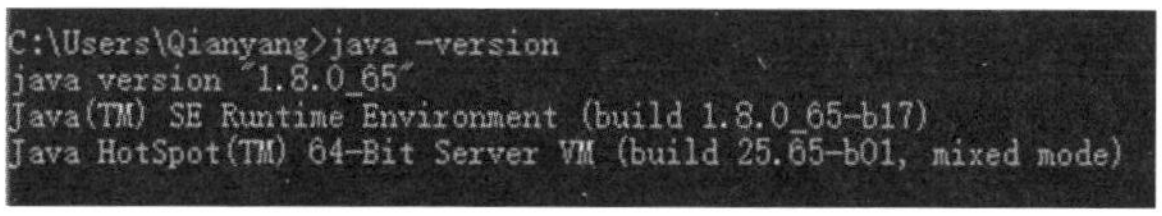

图 2.5　验证配置的环境变量

2.1.2　Eclipse 的下载

开发 Java 项目，需要使用集成开发环境（Integrated Development Environment，IDE）。Eclipse 便是一款免费开源的 Java IDE。图 2.6 所示的 Eclipse IDE for Java EE Developers 为笔者使用的开发工具。

图 2.6　Eclipse IDE for Java EE Developers

首先，下载如图 2.6 所示的 Eclipse，并解压。解压完成后，单击 eclipse.exe 可打开此工具。依次单击 Eclipse 窗口中的 "Window" → "Perferences" → "Java" → "Installed JREs" → "Add" → "Standard VM"，配置开发需要的 JDK，如图 2.7 所示。

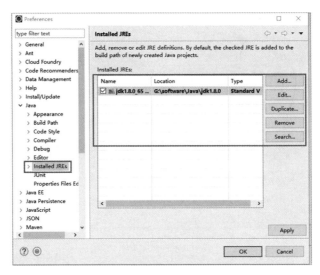

图 2.7　Eclipse 配置 JDK 8

配置完成后，可以新建一个 Java 项目，开发 Java 程序，如图 2.8 所示。

图 2.8　Eclipse 开发 Java 程序

2.2　基本数据类型

在 Java 中，共有 8 种基本数据类型，即 byte、short、int、long、float、double、boolean 和 char。其中，byte、short、int 和 long 为整数类型，float 和 double 为浮点类型，boolean 为逻辑类型，char 为字符类型。

byte 数据类型，内存分配 1 字节，1 字节由 8 位组成，最大值为 2^7-1，最小值为 -2^7，下面为 byte 变量声明方式。

```
byte x = 100;
```

short 数据类型，内存分配 2 字节，占 16 位，最大值为 $2^{15}-1$，最小值为 -2^{15}，下面为 short 变量声明方式。

```
short x = 100;
```

int 数据类型，内存分配 4 字节，占 32 位，最大值为 $2^{31}-1$，最小值为 -2^{31}，下面为 int 变量声明方式。

```
int x = 1000;
```

long 数据类型，内存分配 8 字节，占 64 位，最大值为 $2^{63}-1$，最小值为 -2^{63}，下面为 long 变量声明方式。

```
long x = 100000L;
```

float 数据类型，符合 IEEE754 标准的浮点数，内存分配 4 字节，占 32 位，下面为 float 变量声明方式。

```
float x = 1.5f; 或 float x = 1.5F;
```

double 数据类型，符合 IEEE754 标准的浮点数，内存分配 8 字节，占 64 位，下面为 double 变量声明方式。

```
double x = 100.432;
```

boolean 数据类型，只有两个取值，即 true 和 false，下面为 boolean 变量声明方式。

```
boolean x = true;
```

char 数据类型，即字符类型，内存分配 2 字节，是 1 个单一的 16 位 Unicode 字符，下面为 char 变量声明方式。

```
char x = 'A';
```

2.3 数组

数组是指一组数据的集合，数组中每个数据称为元素或单元。声明数组包括数组类型以及数组名称，如声明一个 int 类型的一维数组。

```
int arr[]; 或 int[] arr; //数组的声明
```

声明数组只是给出了数组的类型以及数组名，要想使用数组，还需要使用 new 语句创建数组，并为数组分配内存，格式如下：

数组名 = new 数据类型[数组元素个数]

例如：

```
arr = new int[4];  //数组创建
```

数组分配空间后，可通过索引的方式初始化数组，索引从 0 开始，如上面的 int 类型数组索引到 3 为止。

```
arr[0] = 1;
arr[1] = 2;
arr[2] = 3;
arr[3] = 4;
```

数组通过索引符访问自身元素，如 arr[0]、arr[1]等，索引从 0 开始。

上述数组的声明、创建和初始化是分开进行的，我们也可以同时进行这三个操作，例如：

```
int[] array = {1, 2, 3, 4};
```

另外，Java 支持多维数组的使用，如 Java 编写文本处理算法时（如主题模型），经常使用二维数组。

再者，数组都有 length 属性，表示操作数组的长度，length 属性只能读取不能修改，如输出某一数组的长度。

```
System.out.println(array.length);
```

在 Java 网络爬虫中，经常涉及数组操作。如采集图片、PDF 和压缩文件时，需要对 byte（字节）类型的数组进行操作；使用 split()方法对字符串类型数据进行分解时，涉及 String 数组的操作。

2.4 条件判断与循环

条件判断语句是 Java 程序中最常见的选择控制结构，共分为三种形式：if、if-else 和 if-else if-else 语句。程序 2-1 为这三种语句的语法格式。

程序 2-1

```
//if语句
if (布尔表达式) {
    //如果表达式为true，执行括号内的语句
}
//if-else语句
if(布尔表达式){
    //如果布尔表达式的值为true
}else{
    //如果布尔表达式的值为false
}
//if-else if-else语句
if(布尔表达式 1){
    //如果布尔表达式 1的值为true执行代码
}else if(布尔表达式 2){
    //如果布尔表达式 2的值为true执行代码
}else if(布尔表达式 3){
    //如果布尔表达式 3的值为true执行代码
}else {
    //如果以上布尔表达式都不为true执行代码
}
```

例如，在网络爬虫中，经常使用 if-else 语句判断网页是否请求成功，如程序 2-2 所示。

程序 2-2

```
if(StatusCode == 200){           //状态码200表示响应成功
    //获取实体内容
    String entity = EntityUtils.toString (response.getEntity(),"gbk");
    //输出实体内容
    System.out.println(entity);
    EntityUtils.consume(response.getEntity());    //消耗实体
}else {
    //关闭HttpEntity的流实体
    EntityUtils.consume(response.getEntity());    //消耗实体
}
```

循环语句是 Java 程序中重要的结构，其功能是反复执行某段程序。Java 中共有三种语句实现循环操作，即 while，do-while 和 for 循环语句。程序 2-3 为这三种语句的语法格式。

程序 2-3

```
//while循环
while( 布尔表达式 ) {
    //循环内容
}
//do-while循环
do {
    //代码语句
}while(布尔表达式);//布尔表达式为终止条件
//for循环
for(初始化; 布尔表达式; 更新) {
    //代码语句
}
```

如在网络爬虫中,使用 while 循环 URL 队列,反复执行 URL 请求以及解析任务,直到队列为空,即所有 URL 采集完成。程序 2-4 为执行的流程。

程序 2-4

```
boolean t = true;
while (t) {
    //如果队列为空,循环结束
    if( queue.isEmpty() ){
        t = false;
    }else {
        //如果队列不为空,则获取数据
    }
}
```

另外,如果采集的页面包含多页,可以通过 for 循环的方式,拼接每页 URL,并将这些 URL 添加到队列中,如程序 2-5 所示。

程序 2-5

```
//队列初始化
Queue<String> queue = new LinkedList<String>();
for(int i = 1; i < 10; i++){
 int page = i;   //爬取的页数
 //拼接URL
 String url = "https://blog.****.net/qy20115549/article/list/" + page;
```

```
        queue.offer(url);   //URL入列
    }
```

2.5　集合

网络爬虫离不开对集合的操作，具体涉及 List、Set、Queue 和 Map 等集合，这些集合都位于 java.util 包。

2.5.1　List 和 Set 集合

List 集合的特征是其元素以线性方式存储,集合中可以存放重复对象。相比而言，Set 集合中的元素不按特定的方式排序，并且没有重复对象。在网络爬虫中，可以使用 List<String> 存储待采集的 URL 列表，如程序 2-6 所示。

程序 2-6

```
//List集合的创建
List<String> urllist = new ArrayList<String>();
//集合元素的添加
urllist.add("https://movie.******.com/subject/27608425");
urllist.add("https://movie.******.com/subject/26968024");
//第一种方式遍历集合
for( String url : urllist ){
    System.out.println(url);
}
//第二种方式遍历集合
for( int i=0; i<urllist.size(); i++ ){
    System.out.println(i+":"+urllist.get(i));
}
//第三种方式遍历集合
Iterator<String> it = urllist.iterator();
while ( it.hasNext() ){
    System.out.println(it.next());
}
```

Set 集合存储不重复的 URL，如程序 2-7 所示。

程序 2-7

```java
//Set集合的初始化
Set<String> set = new HashSet<String>();
set.add("https://movie.******.com/subject/27608425");
set.add("https://movie.******.com/subject/27608425");
set.add("https://movie.******.com/subject/26968024");
//Set集合的遍历
Iterator<String> setIt = set.iterator();
while ( setIt.hasNext() ){
  System.out.println(setIt.next());
}
```

图 2.9 所示为程序 2-7 的执行结果，可以看出 Set 集合自动过滤了重复的元素。

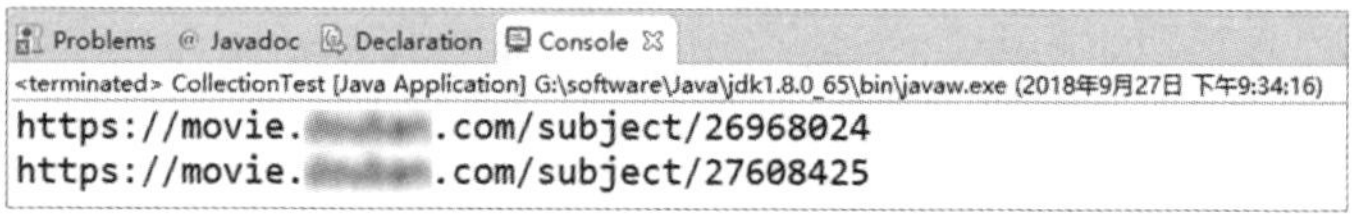

图 2.9　Set 集合案例

2.5.2　Map 集合

Map 集合是一种把键对象和值对象进行映射的集合，它的每个元素都包含一个键对象 key 和值对象 value。其中，键对象不可以重复。Map 集合不仅在网络爬虫中常用，也常在文本挖掘算法编写中使用（如 TF-IDF）。在网络爬虫中，可以使用 Map 集合过滤一些重复数据，但并不建议使用 Map 集合对大规模数据去重过滤，原因是 Map 集合有空间大小的限制。例如，使用网络爬虫采集帖子时，可能遇到置顶帖，而置顶帖在其他页面中也会出现，使用 Map 集合可以过滤已采集的置顶帖。程序 2-8 为 Map 集合使用的方式。

程序 2-8

```java
//Map集合的初始化
 Map<String,Integer> map = new HashMap<String,Integer>();
//值的添加，这里假设是爬虫中的产品id及每个产品id对应的销售量
map.put("jd1515", 100);
map.put("jd1516", 300);
map.put("jd1515", 100);
map.put("jd1517", 200);
map.put("jd1518", 100);
```

```java
//第一种方法遍历Map集合
for (String key : map.keySet()) {
Integer value = map.get(key);
 System.out.println("Key = " + key + ", Value = " + value);
}
//第二种方法遍历Map集合
Iterator<Entry<String, Integer>> entries = map.entrySet().iterator();
while (entries.hasNext()) {
 Entry<String, Integer> entry = entries.next();
 System.out.println("Key = " + entry.getKey() + ", Value = " + entry.getValue());
}
//第三种方法遍历Map集合
for (Entry<String, Integer> entry : map.entrySet()) {
 System.out.println("key= " + entry.getKey() + " and value= " + entry.getValue());
}
```

2.5.3 Queue 集合

Queue(队列)集合使用链表结构存储数据,它只允许在表的前端进行删除操作,而在表的后端进行插入操作,即表的两端操作。Queue 集合常见的操作方法包括添加元素、移除队头元素和判断队列是否为空等,具体使用方法如表 2.1 所示。

表 2.1 Queue 集合常用操作方法

	抛 出 异 常	返回特殊值
添加元素	add(e)	offer(e)
移除元素	remove()	poll()
获取队头元素但不移除队头元素	element()	peek()

在网络爬虫中,Queue 集合常用来存放待采集的 URL,如程序 2-9 所示。

程序 2-9

```java
package com.qian.test;
import java.util.LinkedList;
import java.util.Queue;
public class QueueTest {
 /**
  * 队列常用操作,add()方法和remove()方法在失败的时候会抛出异常(不推荐)
  */
 public static void main(String[] args) {
```

```java
        Queue<String> urlQueue = new LinkedList<String>();
        //添加元素
        urlQueue.offer("https://www.******.com/people/46077896/likes/topic/");
        urlQueue.offer("https://www.******.com/people/1475408/likes/topic");
        urlQueue.offer("https://www.******.com/people/3853295/likes/topic/");
        for(String url : urlQueue){          //遍历队列的元素
            System.out.println(url);
        }
        System.out.println("====================");
        //返回队头元素，并在队列中删除
        System.out.println("第一个url为:" + urlQueue.poll());
        for(String url : urlQueue){
            System.out.println(url);
        }
        System.out.println("====================");
        //获取队头元素但不移除队头元素
        System.out.println("第一个url为:" + urlQueue.element());
        for(String url : urlQueue){
            System.out.println(url);
        }
        System.out.println("====================");
        //获取队头元素但不移除队头元素
        System.out.println("第一个url为:" + urlQueue.peek());
        for(String url : urlQueue){
            System.out.println(url);
        }
        if( urlQueue.isEmpty() ){   //判读队列是否为空
            System.out.println("对列为空！");
        }else {
            System.out.println("队列不为空,包含的元素个数为:" + urlQueue.size());
        }
    }
}
```

执行程序 2-9，会在控制台得到如图 2.10 所示的结果。

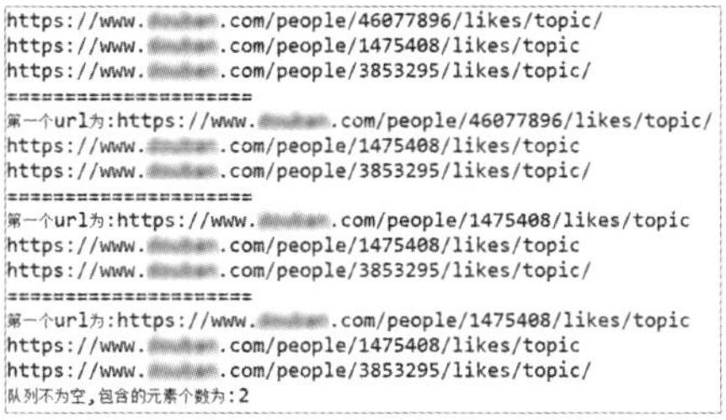

图 2.10 Queue 集合案例运行结果

2.6 对象与类

Java 是一门面向对象的语言，其核心思想是将数据和对数据的操作封装在一起，这里涉及两个重要的概念：对象和类。类封装了一类对象的状态和操作方法，是用来定义对象的模板。对象，是类的具体实例。图 2.11 所示为国外某电商网站，利用网络爬虫采集电商网站商品的 id、product_name（商品名称）、price（商品价格）信息时，每个商品都是一个对象。在编写程序时，需要利用一个类对这些商品信息以及操作方法进行抽象。例如，创建一个 InfoModel 类，类中包含的变量有 id、product_name 和 price，并且使用 private 关键字修饰这些变量，如程序 2-10 所示。

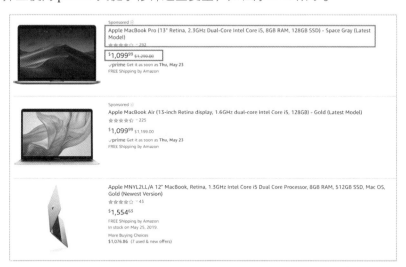

图 2.11 国外某电商网站

在 InfoModel 类中，包含了对每个变量的操作方法，即 set() 与 get() 方法。在采集数据时，设置对象的值可使用 set() 方法，获取对象的值可使用 get() 方法。在程序 2-11 中，使用 InfoModel 类创建了两个对象，封装了两条采集的数据。

程序 2-10

```java
package com.crawler.test;

public class InfoModel {
    private int id;                         //商品id
    private String product_name;            //商品名称
    private double price;                   //商品价格
    public int getId() {
        return id;
    }
    public void setId(int id) {
        this.id = id;
    }
    public String getProduct_name() {
        return product_name;
    }
    public void setProduct_name(String product_name) {
        this.product_name = product_name;
    }
    public double getPrice() {
        return price;
    }
    public void setPrice(double price) {
        this.price = price;
    }
}
```

程序 2-11

```java
package com.crawler.test;
import java.util.ArrayList;
import java.util.List;
public class Test {
    public static void main(String[] args) {
        InfoModel product1 = new InfoModel();   //创建对象
        //对象值的设置
        product1.setId(1);
        product1.setProduct_name("华为p20");
        product1.setPrice(4800.50);
        InfoModel product2 = new InfoModel();   //创建对象
        //对象值的设置
```

```java
        product2.setId(2);
        product2.setProduct_name("华为nova2s");
        product2.setPrice(2300.60);
        //由于所爬数据包含多个,可以封装到集合中进行存储(这里只有两条)
        List<InfoModel> productList = new ArrayList<InfoModel>();
        productList.add(product1);
        productList.add(product2);
        //获取一个对象的产品名称
        System.out.println(productList.get(0).getProduct_name());
    }
}
```

2.7 String 类

Java 使用 java.lang 包中的 String 类来创建字符串变量。另外,String 类提供了操作字符串的一系列方法,表 2.2 列举了常用的一些方法。

表 2.2 String 类操作字符串常用方法

方法	返回值类型	描述
length()	int	获取字符串长度
equals(String s)	boolean	判断两个字符串是否相同
concat(String s)	String	连接两个字符串
contains(String s)	boolean	判断当前字符串是否包含字符串 s
substring(int beginIndex)	String	从 beginIndex 处截取最后所得到的字符串
substring(int beginIndex, int endIndex)	String	从 beginIndex 处到 endIndex 处截取所得到的字符串
indexOf(String s)	int	从字符串头位置开始检索字符串 s,返回首次出现 s 的位置,如果没有检索到返回-1
startsWith(String prefix)	boolean	判断字符串的前缀是否为 prefix
startsWith(String prefix, int toffset)	boolean	判断字符串从指定索引开始的子字符串前缀是否为 prefix,没有检索到返回-1
endsWith(String suffix)	boolean	判断字符串是否以指定后缀 suffix 结束
trim()	String	去除字符串的首尾空格
toLowerCase()	String	将字符串中的所有字符都转换为小写
toUpperCase()	String	将字符串中的所有字符都转换为大写

程序 2-12 演示了表 2.2 中一些方法的使用。

程序 2-12

```java
package com.qian.test;
public class StringTest {
 public static void main(String[] args) {
        String url = " https://www.*****.com/ ";
        String urlTrim = url.trim(); //去除空格字符
        //获取字符串长度
        System.out.println(urlTrim + "\t" + urlTrim.length());
        //转化成大写
        System.out.println("toUpperCase:" + urlTrim.toUpperCase());
        boolean bEqual = urlTrim.equals("www"); //判断字符串是否相同
        boolean bContain = urlTrim.contains("www"); //判断是否包含
        System.out.println("bEqual:" + bEqual + "\t" + "bContain:" + bContain);
        //也可采用+的形式
        String urlConcat = urlTrim.toLowerCase().concat("crawler");
        System.out.println("urlConcat:" + urlConcat);
        //从第二个字符串截取到最后
        String urlSubstring = urlTrim.substring(2, urlTrim.length());
        System.out.println("urlSubstring:" + urlSubstring);
        int urlIndexOf = urlTrim.indexOf("t"); //寻找某字符的位置
        System.out.println("urlIndexOf:" + urlIndexOf);
        //是否以某字符为前缀
        boolean urlStartsWith = urlTrim.startsWith("https");
        //是否以某字符为后缀
        boolean urlEndsWith = urlTrim.endsWith("com/");
        System.out.println("urlStartsWith:" + urlStartsWith + "\t" + "urlEndsWith:"
 + urlEndsWith);
    }
}
```

图 2.12 为程序 2-12 的输出结果。

```
https://www.****.com/   22
toUpperCase:HTTPS://WWW.****.COM/
bEqual:false    bContain:true
urlConcat:https://www.****.com/crawler
urlSubstring:tps://www.****.com/
urlIndexOf:1
urlStartsWith:true      urlEndsWith:true
```

图 2.12　String 类的使用

另外，调用 java.lang 包的其他类中的方法，可以实现 String 数据类型与其他数据类型的转换。如采集数据时，经常需要将字符串类型转化成整型和 Double 类型等（见程序 2-13）。

程序 2-13

```
String sumPage = "30";    //例如，某论坛的帖子总页数为 30
int sumPageParse = Integer.parseInt(sumPage);
String price = "1299.8";  //例如，某产品的价格为 1299.8
double priceParse = Double.parseDouble(price);
System.out.println(sumPageParse + "\t" + priceParse);
```

2.8 日期和时间处理

java.util 包提供的处理日期和时间的类有 SimpleDateFormat、DateFormat、Date 以及 Calendar。其中，SimpleDateFormat 和 DateFormat 类用来实现日期和时间的格式化，常使用的方法有 format(Date date) 和 parse(String source)；Date 类主要处理指定格式的时间；Calendar 类主要用于年月日类型数据的转换。

在采集数据时，不同网站的时间使用格式可能不同。而不同的时间格式，会为数据存储以及数据处理带来一定的困难。例如，图 2.13 所示为某汽车论坛中的时间使用格式，包含 "yyyy-MM-dd" 和 "yyyy-MM-dd HH:mm" 两种格式；图 2.14 所示为某新闻网站中的时间使用格式为 "yyyy-MM-dd HH:mm:ss"。

针对汽车论坛中的 "yyyy-MM-dd" 和 "yyyy-MM-dd HH:mm" 格式，可以统一转化成 "yyyy-MM-dd HH:mm:ss" 格式，以方便数据存储以及后期数据处理，如程序 2-14 所示。在程序 2-14 中，parseStringTime() 方法的作用是将字符串类型的时间标准化成指定格式的时间，这里通过调用 DateFormat 类中的 parse() 方法和 format() 方法来实现。

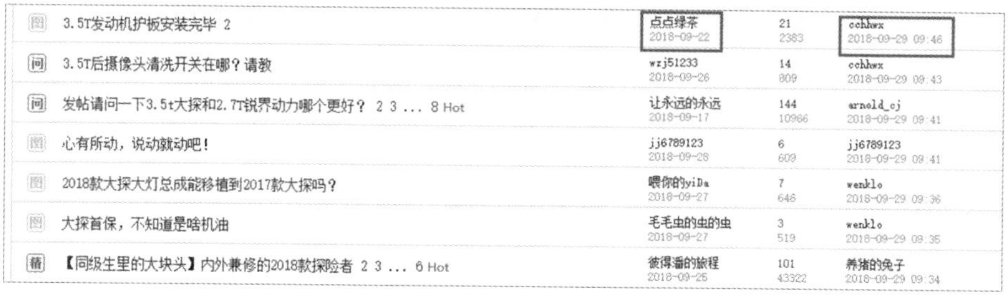

图 2.13　某汽车论坛中的时间使用格式

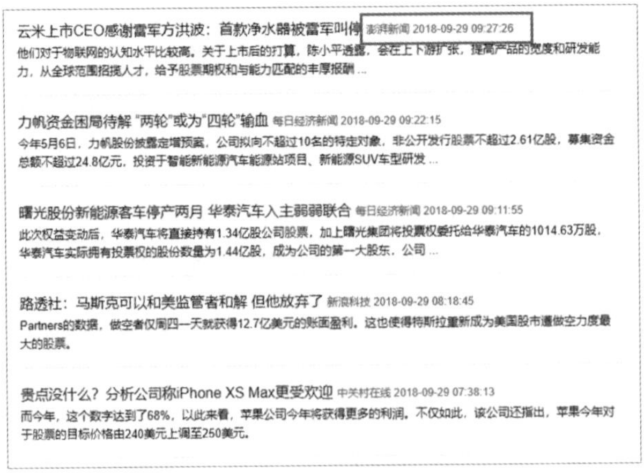

图 2.14　某新闻网站中的时间使用格式

程序 2-14

```java
package com.qian.test;
import java.text.ParseException;
import java.text.SimpleDateFormat;
import java.util.Date;
public class TimeTest {
  public static void main(String[] args) {
      System.out.println(parseStringTime("2016-05-19 19:17",
            "yyyy-MM-dd HH:mm","yyyy-MM-dd HH:mm:ss"));
      System.out.println(parseStringTime("2018-06-19",
            "yyyy-MM-dd","yyyy-MM-dd HH:mm:ss"));
  }
  /**
   * 字符型时间格式标准化方法
   * @param inputTime(输入的字符串时间), inputTimeFormat(输入的格式),
outTimeFormat(输出的格式)
   * @return 转化后的时间(字符串)
   */
  public static String parseStringTime(String inputTime,String inputTimeFormat,
        String outTimeFormat){
    String outputDate = null;
    try {
        //日期格式化及解析时间
        Date inputDate = new SimpleDateFormat(inputTimeFormat).parse(inputTime);
```

```
        //转化成新的形式的字符串
        outputDate = new SimpleDateFormat(outTimeFormat).format
(inputDate);
    } catch (ParseException e) {
        e.printStackTrace();
    }
    return outputDate;
    }
}
```

图 2.15 所示为程序 2-14 的输出结果。

```
2016-05-19 19:17:00
2018-06-19 00:00:00
```

图 2.15 格式化时间

另外，采集数据时，通常需要记录每一条数据的采集时间。图 2.16 所示为采集的汽车销量数据，字段 craw_time 表示采集每条记录的时间信息。程序 2-15 给出了获取当前时间的一个通用方法。在该方法中，使用了 Date 类的无参构造方法创建了一个 Date 对象，进而获取当前时间。

程序 2-15

```
//获取系统的当前时间
public static String GetNowDate(String formate){
    String temp_str = "";
    Date dt = new Date();
    SimpleDateFormat sdf = new SimpleDateFormat(formate);
    temp_str = sdf.format(dt);
    return temp_str;
}
```

car_id	date	salesnum	craw_time
1001	2007-01-01	14834	2016-12-08 13:57:36
1001	2007-02-01	9687	2016-12-08 13:57:36
1001	2007-03-01	18173	2016-12-08 13:57:36
1001	2007-04-01	18508	2016-12-08 13:57:36
1001	2007-05-01	19710	2016-12-08 13:57:36
1001	2007-06-01	20311	2016-12-08 13:57:36
1001	2007-07-01	17516	2016-12-08 13:57:36

图 2.16 采集的汽车销量数据

再者，一些网站的时间格式为 UNIX 时间戳（即从 1970 年 1 月 1 日开始经过的秒数）。图 2.17 所示为国外某艺术品网站，基于浏览器抓包（下一章中讲解），发现该网页的时间格式为 UNIX 时间戳，如图 2.18 所示。为将 UNIX 时间戳转化为指定格式的时间，需要构造一个通用的方法，如程序 2-16 所示。

程序 2-16

```java
//将UNIX时间戳转化成指定格式的时间
public static String TimeStampToDate(String timestampString,
String formats) {
    Long timestamp = Long.parseLong(timestampString) * 1000;
    String date = new SimpleDateFormat(formats,
        Locale.CHINA).format(new Date(timestamp));
    return date;
}
```

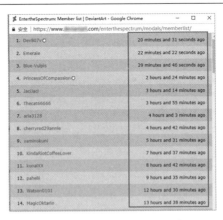

图 2.17　国外某艺术品网站

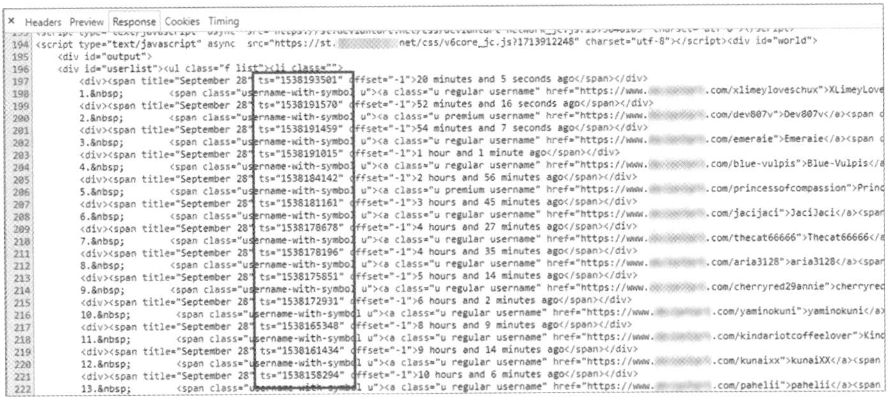

图 2.18　网页中的 UNIX 时间戳

2.9　正则表达式

正则表达式指由普通字符（如英文字母/数字等）以及特殊字符（如元字符"\D"）

组成的字符串模式。正则表达式可以理解成一套模板,以这套模板可以匹配字符串。例如,从下面 HTML 片段中提取用户的 id(75975500)。

```
<a href="//i.********.com.cn/75975500" target="_blank" class="linkblack">尊少来自沈阳
```

第 5 章会介绍一些 HTML 网页解析工具,如上面的 HTML 片段,使用 Jsoup 可直接解析得到字符串 "//i.********.com.cn/75975500"。针对这个字符串,调用 String 类中的 replaceAll(String regex, String replacement)方法,可成功剔除非数字字符(元字符 "\D" 表示非数字字符),进而提取用户 id,如程序 2-17 所示。其中, replaceAll() 方法的作用是替换字符串中的某段内容,参数 regex 为正则表达式,replacement 为替换的内容。

程序 2-17

```
//使用 Jsoup 解析得到的 URL 片段
String url = "//i.********.com.cn/75975500";
String user_id = url.replaceAll("\\D", "");   //取代所有的非数字字符
System.out.println(user_id);   //输出的结果即为 75975500
```

另外,分解某字符串也常使用正则表达式,如将字符串"正则表达式-CSDN 博客"分割为 "正则表达式" 和 "CSDN 博客" 两个字符串,可利用 String 类中的 split(regex) 方法对原字符串进行操作,如程序 2-18 所示。其中,正则表达式 "\\p{Punct}" 匹配的是标点符号。

程序 2-18

```
String content = "正则表达式-CSDN 博客";
//以标点符号进行切割
String[] contentRegex = content.split("\\p{Punct}");
//输出结果为"正则表达式"
System.out.println(contentRegex[0]);
```

表 2.3 列举了常用元字符以及它们的正则表达式写法。

表 2.3 常用元字符及其含义

元 字 符	正则表达式写法	含 义
\d	\\d	代表 0~9 中的任意数字
\D	\\D	代表任何一个非数字字符
\S	\\S	代表非空格类字符

续表

元 字 符	正则表达式写法	含 义
\s	\\s	代表空格类字符
\p{Lower}	\\p{Lower}	代表小写英文字母
\p{Upper}	\\p{Upper}	代表大写英文字母
\p{Punct}	\\p{Punct}	代表标点符号
\p{Blank}	\\p{Blank}	代表空格或制表符（\t）

在正则表达式中，常用方括号括起若干字符表示一个元字符，该元字符匹配的是括号内的任意一个字符。例如，String regex = "[abc]123" 匹配的是 "a123"，"b123"，"c123"。表 2.4 列举了一些包含方括号的元字符。

表 2.4 包含方括号的元字符

写　　法	含　　义
[abc]	a 或 b 或 c
[^abc]	除去 abc 的任何字符
[a-z]	表示 a~z 中的任何一个字母
[a-zA-Z]	所有英文字母
[1-9]	表示 1~9 中的任何一个数字
[a-d1-3]	字母 a~d 和数字 1~3

程序 2-19 演示了表 2.4 中元字符的使用，图 2.19 所示为该程序的输出结果。

程序 2-19

```
package com.qian.test;
public class RegexTest1 {
  public static void main(String[] args) {
      String str1 = "a1b2c3dAZ4";
      String strReplace1 = str1.replaceAll("[abc]", "");
      System.out.println("使用元字符[abc]匹配的结果为:" + strReplace1);
      String strReplace2 = str1.replaceAll("[^abc]", "");
      System.out.println("使用元字符[^abc]匹配的结果为:" + strReplace2);
      String strReplace3 = str1.replaceAll("[a-zA-Z]", "");
      System.out.println("使用元字符[a-zA-Z]匹配的结果为:" + strReplace3);
      String strReplace4 = str1.replaceAll("[1-9]", "");
      System.out.println("使用元字符[1-9]匹配的结果为:" + strReplace4);
      String strReplace5 = str1.replaceAll("[1-3]", "");
      System.out.println("使用元字符[1-3]匹配的结果为:" + strReplace5);
      String strReplace6 = str1.replaceAll("[a-d1-3]", "");
      System.out.println("使用元字符[a-d1-3]匹配的结果为:" + strReplace6);
```

```
    }
}
```

```
使用元字符[abc]匹配的结果为:123dAZ4
使用元字符[^abc]匹配的结果为:abc
使用元字符[a-zA-Z]匹配的结果为:1234
使用元字符[1-9]匹配的结果为:abcdAZ
使用元字符[1-3]匹配的结果为:abcdAZ4
使用元字符[a-d1-3]匹配的结果为:AZ4
```

图 2.19　程序 2-19 的输出结果

为了适应匹配的不确定性，正则表达式支持限定符的概念。限定符定义了某些元素可以出现的频次。例如，X{n,m}表示 X 出现 n 到 m 次的字符都可以匹配，使用"ac{1,3}"可以匹配"ac"、"acc"和"accc"。表 2.5 列举了常用的限定符。

表 2.5　常用限定符

写　　法	含　　义
X{n}	X 确定出现 n 次，如"a{2}"不能匹配"cab"中的"a"，但能匹配"caab"中的两个"a"
X{n,}	X 至少出现 n 次，如"a{2,}"不能匹配"cab"中的"a"，但能匹配"caaab"中的 3 个"a"
X{n,m}	X 出现 n 到 m 次，如"ac{1,3}"可以匹配"ac"、"acc"和"accc"
X?	X 出现 0 次或 1 次，如"ac[12]?"匹配的是"ac"、"ac1"和"ac2"
X*	X 出现 0 次或多次，如"ac*"能匹配"a"、"ac"和"acc"等
X+	表示 X 出现 1 次或多次，如"ac+"能匹配"ac"、"acc"和"accc"等

2.10　Maven 工程的创建

Maven 是由 Apache 软件基金会提供的一款工具，用于项目管理及自动构建。在构建一个 Java 工程时，需要用到许多 jar 包，如操作数据库需要用到 mysql-connector-java 以及其相关依赖 jar 包。而 Maven 可以很方便地管理 Java 工程中所需的 jar 包（如自动下载 jar 包及依赖 jar 包、导入 jar 包、移除 jar 包等）。

Maven 使用项目对象模型（Project Object Model，POM）来配置，项目对象模型存储在 pom.xml 文件中。在 Eclipse 中，单击"File→New→Other→Maven→Maven Project"，可以创建 Maven 工程，如图 2.20 所示。在创建过程中，涉及两个必填项，即 Group Id 和 Artifact Id。其中，Group Id 为项目组织唯一标识符，实际对应的是 Java 项目的包（package）的名称；Artifact Id 指项目的唯一标识符，对应的是 Java 项目名称，如图 2.21 所示，创建的项目名称为 Test，包的名称为 com.qian.test。图 2.22 所示为项目 Test 具体的目录结构，可以看到 pom.xml 文件在 Test 项目的根目录下，Maven Dependencies 存放着 Maven 工程管理的 jar 包。

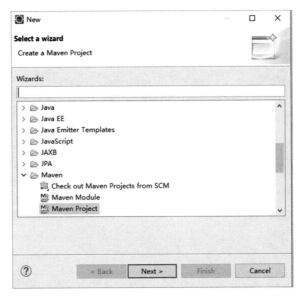

图 2.20　在 Eclipse 中创建 Maven 工程

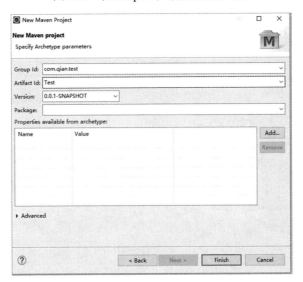

图 2.21　Group Id 和 Artifact Id 的填写

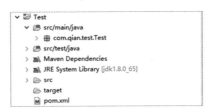

图 2.22　项目 Test 的目录结构

程序 2-20 为项目 Test 中 pom.xml 文件的内容。其中，group Id 和 artifact Id 前面已经有过介绍，version 为工程的版本号；packaging 为打包机制，如 jar、war 等；properties 定义了一些常量，以方便其他地方直接引用；dependencies 用来配置工程需要引入的 jar 包，其中每个 dependency 都对应一个 jar 包。在引入 dependency 中的某个 jar 时，也会将其依赖的 jar 包引入。

程序 2-20

```
<project xmlns="http://maven.******.org/POM/4.0.0" xmlns:xsi=
"http://www.**.org/2001/XMLSchema-instance"
    xsi:schemaLocation="http://maven.******.org/POM/4.0.0
http://maven.******.org/xsd/maven-4.0.0.xsd">
    <modelVersion>4.0.0</modelVersion>
    <groupId>com.qian.test</groupId>
    <artifactId>Test</artifactId>
    <version>0.0.1-SNAPSHOT</version>
    <packaging>jar</packaging>
    <name>Test</name>
    <url>http://maven.******.org</url>
    <properties>
        <project.build.sourceEncoding>UTF-8</project.build.sourceEncoding>
    </properties>
    <dependencies>
      <dependency>
        <groupId>junit</groupId>
        <artifactId>junit</artifactId>
        <version>3.8.1</version>
        <scope>test</scope>
      </dependency>
    </dependencies>
</project>
```

如在网络爬虫中，为引入网页请求工具 HttpClient、网页解析工具 Jsoup 和数据库连接工具 mysql-connector-java 相关的 jar 包，可在 dependencies 中添加相应的 dependency，如程序 2-21 所示。加入 dependency 后，可使用快捷键 Ctrl+S 保存 pom.xml 文件，项目会自动下载或添加相应 jar 包及其依赖 jar 包，如图 2.23 所示。另外，通过 Maven Dependencies，可以查看每个 jar 包的存储路径。

程序 2-21

```xml
<dependency>
        <groupId>mysql</groupId>
        <artifactId>mysql-connector-java</artifactId>
        <version>5.1.35</version>
</dependency>
<dependency>
        <groupId>org.jsoup</groupId>
        <artifactId>jsoup</artifactId>
        <version> 1.8.2</version>
</dependency>
<dependency>
        <groupId>org.apache.httpcomponents</groupId>
        <artifactId>httpclient </artifactId>
        <version>  4.2.3</version>
</dependency>
```

图 2.23　项目下载或添加 jar 包及其依赖 jar 包

最后，推荐一个非常有用的 Maven 仓库搜索网站 Maven Repository。基于该网站的搜索功能，可以快速查找和配置 Java 项目所需 jar 包及其依赖 jar 包。图 2.24 所示为搜索关键词 "mysql" 对应的页面内容，其中，第一个 "MySQL Connector/J" 的使用量为 3026。单击 "MySQL Connector/J"，可以看到不同版本的 jar 包，之后，单击符合自身需求的 jar 包版本（如 5.1.47 版本），会看到 jar 包对应的 dependency 写法以及依赖 jar 包，如图 2.25 所示。

第 2 章 网络爬虫涉及的 Java 基础知识   33

图 2.24　Maven Repository 搜索页面

图 2.25　dependency 写法及依赖 jar 包

2.11　log4j 的使用

  日志是程序的重要组成部分，在程序中添加日志记录，可以很方便地记录代码变化情况、跟踪代码运行轨迹、发掘代码错误位置并调试。Apache 软件基金会下的开源项目 log4j 便是一个功能强大的日志框架，其可以将日志信息输出到控制台和指定文件中,相比直接使用 System.out.println()将信息打印在控制台更具优势。

使用 log4j 前,需要使用 Maven 下载 jar 包,即创建一个 Maven 工程,并在 pom.xml 文件中添加 log4j 对应的 dependency。

```
<dependency>
 <groupId>log4j</groupId>
 <artifactId>log4j</artifactId>
 <version>1.2.17</version>
</dependency>
```

接着,在 src/java/main 目录下创建 log4j.properties,用于配置日志信息,如图 2.26 所示。如果创建的是普通 Java 工程,可在 src 目录下创建 log4j.properties。

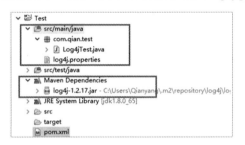

图 2.26 创建 log4j.properties

下面进行配置 log4j.properties 文件,其分为以下几个步骤。

步骤 1,配置根 Logger,使用语法如下所示。

```
log4j.rootLogger = [ level ] , appenderName, appenderName, …
```

在上述语法中,level 为日志级别,优先级从低到高依次为 DEBUG(调试)、INFO(信息)、WARN(警告)、ERROR(错误)和 FATAL(重大错误,会导致程序直接中断)。这 5 个级别之间是包含关系,如定义 level 为 INFO 级别,则大于及等于 INFO 级别的日志信息都可以输出,而低于 INFO 级别的 DEBUG 信息将无法输出。

appenderName 为日志输出目的地名称(可同时定义多个),如用 stdout 表示标准控制输出(即控制台)、用 D 表示 DEBUG 信息的输出、用 E 表示 ERROR 信息的输出。

步骤 2,配置日志的输出目的地 Appender,使用语法如下所示。

```
log4j.appender.appenderName = fully.qualified.name.of.appender.class
log4j.appender.appenderName.option1 = value1
  …
log4j.appender.appenderName.optionN = valueN
```

在上述语法中，appenderName 为根目录中定义的输出目的地名称，log4j 提供的 appender 有 ConsoleAppender、FileAppender、DailyRollingFileAppender、RollingFileAppender 和 WriterAppender。由于 WriterAppender 不能直接配置使用，所以使用更多的是前面 4 种，具体如下所示。

```
org.apache.log4j.ConsoleAppender（控制台）
org.apache.log4j.FileAppender（文件）
org.apache.log4j.DailyRollingFileAppender（每天产生一个日志文件）
org.apache.log4j.RollingFileAppender（文件大小到达指定尺寸的时候产生
一个新的文件）
org.apache.log4j.WriterAppender（将日志信息以流格式发送到任意指定的地方）
```

ConsoleAppender（控制台）的 option 包括如下。
- Threshold = DEBUG：指定日志消息的输出最低级别，默认为 DEBUG。
- ImmediateFlush = true：默认值是 true，表示所有消息都会被立即输出。
- Target = System.err：使用 System.err 在控制台输出，默认情况下是 System.out。

FileAppender（文件）的 option 包括如下。
- Threshold = DEBUG：指定日志消息的输出最低级别，默认为 DEBUG。
- ImmediateFlush = true：默认值是 true，表示所有消息都会被立即输出。
- File = E:\logs.txt：指定消息输出到 E 盘中的 logs.txt 文件。
- Append = false：默认值是 true，即以追加的方式将消息增加到指定文件中，false 指将消息覆盖指定的文件内容。

DailyRollingFileAppender（每天产生一个日志文件）的 option 包括如下。
- Threshold = DEBUG：指定日志消息的输出最低级别，默认为 DEBUG。
- ImmediateFlush = true：默认值是 true，表示所有消息都会被立即输出。
- File = E:\logs.txt：指定消息输出到 E 盘中的 logs.txt 文件。
- Append = false：默认值是 true，即以追加的方式将消息增加到指定文件中，false 指将消息覆盖指定的文件内容。
- DatePattern = '.'yyyy-ww：每周滚动一次文件，即每周产生一个新的文件。用户可以指定按月、周、天、时和分滚动文件。

RollingFileAppender（文件大小达到指定大小的时候产生一个新的文件）的 option 包括如下。
- Threshold = DEBUG：指定日志消息的输出最低级别，默认为 DEBUG。
- ImmediateFlush = true：默认值是 true，表示所有消息都会被立即输出。

- File= E:\logs.txt：指定消息输出到 E 盘中的 logs.txt 文件。
- Append = false：默认值是 true，即以追加的方式将消息增加到指定文件中，false 指将消息覆盖指定的文件内容。
- MaxFileSize = 100KB：其后缀可以是 KB、MB 或 GB。在日志文件达到设置值时，将会产生新文件，即将原来的内容移到 logs.txt.1 文件。
- MaxBackupIndex = 2：指定可以产生的滚动文件的最大数。

步骤 3，配置日志信息布局 layout，使用语法如下所示。

```
log4j.appender.appenderName.layout = fully.qualified.name.of.layout.class
log4j.appender.appenderName.layout.option1 = value1
…
log4j.appender.appenderName.layout.optionN = valueN
```

log4j 提供的 layout 有 HTMLLayout、PatternLayout、SimpleLayout 和 TTCCLayout，较为常用的是 PatternLayout。

```
org.apache.log4j.HTMLLayout（HTML表格形式布局）
org.apache.log4j.PatternLayout（灵活地指定布局模式）
org.apache.log4j.SimpleLayout（包含日志信息的级别和信息字符串）
org.apache.log4j.TTCCLayout（包含日志产生的时间、线程、类别等信息）
```

HTMLLayout 的 option 包括如下。

- LocationInfo = true：默认值是 false，输出 java 文件名称和行号。
- Title = Test_WARN：默认值是 Log4J Log Messages，这里为 Test_WARN。

PatternLayout 的 option 包括如下。

- ConversionPattern = %m%n：以指定信息格式输出，格式如下所示。

%p：输出优先级，即 DEBUG、INFO、WARN、ERROR、FATAL。

%r：输出自应用启动到输出该 log 信息耗费的毫秒数。

%c：输出所属的类目，通常就是所在类的全名。

%n：输出一个回车换行符。

%d：输出日志时间点的日期或时间，默认格式为 ISO8601，也可以在其后指定格式，比如，%d{yyyy-MM-dd HH\:mm\:ss,SSS}。

%l：输出日志事件的发生位置，包括类目名、发生的线程，以及在代码中的行数。

%t：输出产生该日志事件的线程名。

程序 2-22 所示为 log4j.properties 配置的一个案例。

程序 2-22

```
###配置根 Logger###
log4j.rootLogger = debug,stdout,D,E,W
### 输出信息到控制台 ###
log4j.appender.stdout = org.apache.log4j.ConsoleAppender
log4j.appender.stdout.Target = System.out
log4j.appender.stdout.layout = org.apache.log4j.PatternLayout
log4j.appender.stdout.layout.ConversionPattern=%p[%d{yyyy-MM-dd HH\:mm\:ss,SSS}] [%t] %C.%M(%L) | %m%n
### 输出DEBUG 级别以上的日志到=E://logs/error.log ###
log4j.appender.D = org.apache.log4j.DailyRollingFileAppender
log4j.appender.D.File = E:/logs/log.log
log4j.appender.D.Append = true
log4j.appender.D.Threshold = DEBUG
log4j.appender.D.layout = org.apache.log4j.PatternLayout
log4j.appender.D.layout.ConversionPattern = %p[%d{yyyy-MM-dd HH\:mm\:ss,SSS}] [%t] %C.%M(%L) | %m%n
### 输出ERROR 级别以上的日志到=E://logs/error.log ###
log4j.appender.E = org.apache.log4j.DailyRollingFileAppender
log4j.appender.E.File = E:/logs/error.log
log4j.appender.E.Append = true
log4j.appender.E.Threshold = ERROR
log4j.appender.E.layout = org.apache.log4j.HTMLLayout
log4j.appender.E.layout.LocationInfo = true
log4j.appender.E.layout.Title = Test_ERROR
### 输出INFO 级别以上的日志到=E://logs/warn.log   ###
log4j.appender.W = org.apache.log4j.RollingFileAppender
log4j.appender.W.File = E:/logs/warn.log
log4j.appender.W.Append = true
log4j.appender.W.Threshold = WARN
log4j.appender.W.MaxFileSize = 2KB
log4j.appender.W.layout = org.apache.log4j.HTMLLayout
log4j.appender.W.layout.LocationInfo = true
log4j.appender.W.layout.Title = Test_WARN
```

基于程序 2-22 配置的 log4j.properties，下面构建 log4j 使用的案例，如程序 2-23 所示。

程序 2-23

```
package com.qian.test;
import org.apache.log4j.Logger;
public class Log4jTest {
  static final Logger logger = Logger.getLogger(Log4jTest.class);
  public static void main(String[] args) {
    System.out.println("hello");  //控制台输出
    //日志信息
    logger.info("hello world");
    logger.debug("This is debug message.");
    logger.warn("This is warn message.");
    logger.error("This is error message.");
  }
}
```

在 Eclipse 中执行程序 2-23，控制台会输出如下结果。

```
hello
 INFO[2018-10-02 10:21:07,633] [main] com.qian.test.Log4jTest.main(8) | hello world
DEBUG[2018-10-02 10:21:07,635] [main] com.qian.test.Log4jTest.main(9) | This is debug message.
 WARN[2018-10-02 10:21:07,635] [main] com.qian.test.Log4jTest.main(10) | This is warn message.
ERROR[2018-10-02 10:21:07,636] [main] com.qian.test.Log4jTest.main(11) | This is error message.
```

从结果中看出 System.out.println()输出的只是括号中的内容，而日志可以输出更多的信息，如程序执行的具体时间、程序执行的位置以及输出内容等。同时，在目录 logs 下，自动生成了三个日志文件，如图 2.27 所示。

图 2.27　日志文件

由于 layout 的 ConversionPattern 的设置与控制台日志输出的设置相同，所以文件夹下的 log.log 的日志信息与控制台中的日志信息相同，日志信息如下所示。

```
    INFO[2018-10-02 10:21:07,633] [main] com.qian.test.Log4jTest.main
(8) | hello world
    DEBUG[2018-10-02 10:21:07,635] [main] com.qian.test.Log4jTest.
main(9) | This is debug message.
    WARN[2018-10-02 10:21:07,635] [main] com.qian.test.Log4jTest.main
(10) | This is warn message.
    ERROR[2018-10-02 10:21:07,636] [main] com.qian.test.Log4jTest.
main(11) | This is error message.
```

而 error.log 和 warn.log 中的日志内容均以 HTML 形式布局，下面给出了 error.log 中的日志内容，包括 Time、Thread、Level、Category、File:Line 和 Message 方面的信息。同时，从下面的日志内容也可以发现该文件是缺少</body>和</html>的。

```
    <!DOCTYPE HTML PUBLIC "-//W3C//DTD HTML 4.01 Transitional//EN"
"http://www.**.org/TR/html4/loose.dtd">
    <html>

    <head>
        <title>Test_ERROR</title>
        <style type="text/css">
        <!-- body, table {font-family: arial,sans-serif; font-size:
x-small;} th {background: #336699; color: #FFFFFF; text-align: left;}
--></style>
    </head>

    <body bgcolor="#FFFFFF" topmargin="6" leftmargin="6">
        <hr size="1" noshade>Log session start time Tue Oct 02 10:21:07
CST 2018
        <br>
        <br>
        <table cellspacing="0" cellpadding="4" border="1" bordercolor
="#224466" width="100%">
        <tr>
        <th>Time</th>
        <th>Thread</th>
        <th>Level</th>
        <th>Category</th>
        <th>File:Line</th>
        <th>Message</th></tr>
        <tr>
        <td>3</td>
        <td title="main thread"
```

```
>main</td>
        <td title="Level">
          <font color="#993300">
            <strong>ERROR</strong></font>
        </td>
        <td title="com.qian.test.Log4jTest category">com.qian.
test.Log4jTest</td>
        <td>Log4jTest.java:11</td>
        <td title="Message">This is error message.</td></tr>
```

2.12 本章小结

本章着重介绍了网络爬虫涉及的 Java 基础知识,包括开发环境的搭建、数据类型、数组等方面的内容。当然,网络爬虫还涉及其他方面的 Java 知识,如关键字、继承与接口、异常、输入流与输出流和多线程等。在后续章节中,会通过具体的网络爬虫介绍相关 Java 知识点的使用。

第 3 章

HTTP 协议基础与网络抓包

3.1 HTTP 协议简介

用户在 Web 浏览器（下文统称为客户端）请求某个 URL 时，该请求会被发送到 Web 服务器，Web 服务器接收到客户端的请求后，将响应请求，并向客户端传送数据。为保证数据在客户端和 Web 服务器之间传输信息的可靠性，客户端和 Web 服务器必须遵守一定的标准或规则，其中最为重要的便是 HTTP（Hyper Text Transfer Protocol）协议。

HTTP 协议，即超文本传输协议，是 Web 系统最核心的内容，用于从 Web 服务器传输数据到客户端。图 3.1 所示为基于 HTTP 协议的数据传输，其传输的数据类型有数百个，以下列举了部分数据类型。

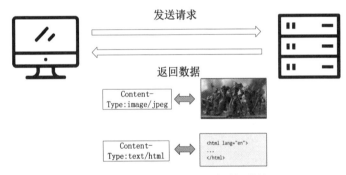

图 3.1 基于 HTTP 协议的数据传输

- text/html：HTML 格式的文本文档。
- image/jpeg：JPEG 格式的图片。
- image/png：PNG 格式的图片。
- image/webp：无损格式的图片。
- image/gif：GIF 格式的图片。
- text/plain：普通的 ASCII 文本文档。
- application/json：JSON 格式的内容。
- video/mp4：MP4 格式的视频。

- video/quicktime：Apple 的 QuickTime 视频（MOV 格式的视频）。
- video/x-msvideo：AVI 格式的视频。
- video/x-flv：FLV 格式的视频。

HTTP 是应用层协议，其数据在网络中传输需要依赖于 TCP/IP 协议。TCP（Transmission Control Protocol），即传输控制协议，用于保证数据在两台主机之间传输的可靠性，TCP 实行"顺序控制"（数据会按照发送的顺序到达）和"重发控制"（传输数据时，为每个数据包启动一个超时定时器，如果定时器在超时之前，接收方确认了信息，则释放数据包占用的缓存区，否则，发送方就重新发送这个数据包）。IP（Internet Protocol），即网际协议，负责将数据包从源发送到最终的目标计算机，但不提供可靠性传输，也不具备"重发控制"。如图 3.2 所示，TCP/IP 协议通常被认为是具有 4 个层次的系统，即数据链路层、网络层、传输层和应用层，每个层次负责不同的功能。

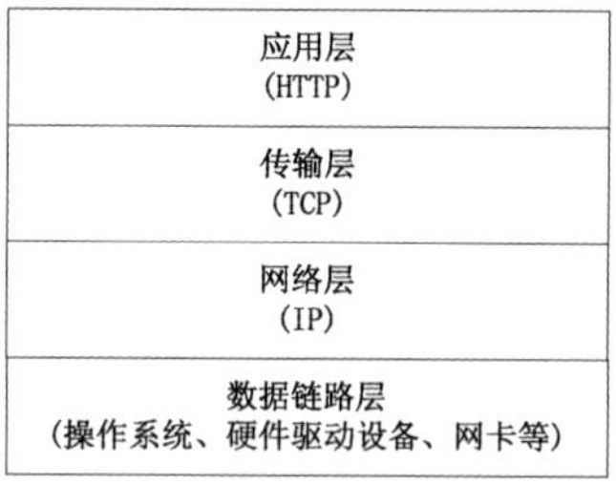

应用层 （HTTP）
传输层 （TCP）
网络层 （IP）
数据链路层 （操作系统、硬件驱动设备、网卡等）

图 3.2 TCP/IP 协议的四个层次

- 应用层：该层位于计算机网络体系结构的最上层，其他三层为其服务。日常开发以及使用的应用软件便处于应用层，如使用浏览器访问网页。TCP/IP 协议的应用架构多采用客户端/服务端（Client/Server）模型，即客户端（如浏览器）向服务端（服务器）发送请求，服务端向客户端传送数据。因此，可以看出 HTTP 协议处于应用层。
- 传送层：该层用于保证客户端与服务端之间的数据传输。
- 网络层：该层用于选择合适的网间路由和交换节点，处理网络中流动的数据包（数据传输的最小单位）。
- 数据链路层：该层又称为网络接口层，用于处理连接网络的硬件部分，包括操作系统、硬件驱动设备、网卡和光纤等。

3.2 URL

在进行网络通信前，需要通过 URL 建立客户端与服务器之间的连接。URL

（Uniform Resource Locator），即统一资源定位符，是统一资源标识符 URI（Uniform Resource Identifier）的子集。URL 描述了请求资源在某个特定服务器的位置信息。基于 URL，可以精确地定位到网络资源。图 3.3 展示了如何通过 URL 访问某服务器的图片资源。

图 3.3　URL 定位资源

以图 3.3 所示 URL 为例，介绍 URL 的组成部分。

- 协议：该 URL 使用的协议为 HTTP，在 HTTP 后面跟 "://"。有很多网站，从安全角度考虑，使用的是安全超文本传输协议（HyperText Transfer Protocol Secure，HTTPS）。HTTPS 在 HTTP 的基础上加入了安全套接层（Secure Sockets Layer，SSL）。SSL 主要用于客户端与服务器之间的身份认证和数据加密传输，其支持使用 X509 数字认证。如使用 HttpClient 处理一些 HTTPS 对应的 URL 时，需要创建定制的 SSL 连接（第 7 章的实战案例中有详细讲解）。
- 域名：图 3.3 中 URL 的域名为 www.*****.com。如果没有域名，则必须有客户端可以访问到的主机 IP 地址。例如，在如图 3.4 所示命令提示符 cmd 中，使用 ping www.*****.com 获取该网站的 IP 地址，在浏览器中输入该 IP 地址也可访问网页。
- 端口：位于域名的后面，域名和端口采用 ":" 分隔。如果使用默认端口号 80，则 URL 可以省略端口信息，如图 3.3 中的 URL。
- 路径：由多个 "/" 隔开的字符串组成，表示主机上的一个目录或文件地址。如图 3.3 中的 "/uploads/allimg/170501/1_05011010139161.jpg"。
- 参数：以 "?" 开始，采用 name=value 的格式。URL 中的参数可以有多个，参数之间用一个 "&" 隔开。如在某购物网站中搜索关键词 "手机"，得到下面一串 URL，其中参数为"keyword=手机&enc=utf-8&wq=手机&pvid=101d1b98589b41d3a456c008fcbf79cc"。

```
https://search.**.com/Search?keyword=%E6%89%8B%E6%9C%BA&enc=utf-8
&wq=%E6%89%8B%E6%9C%BA&pvid=101d1b98589b41d3a456c008fcbf79cc
```

对于浏览器（客户端）获取服务器资源的详细步骤如下所示。

步骤1，浏览器从输入的 URL 中解析出服务器的域名和端口号（如果没有端口号，则为 80）。

步骤2，浏览器将服务器的域名转化为服务器的 IP 地址。

步骤3，基于服务器的 IP 地址及端口号，建立浏览器与服务器的 TCP 连接。

步骤4，浏览器向服务器发送 HTTP 请求报文。

步骤5，基于浏览器请求内容，服务器向浏览器返回相应的 HTTP 响应报文。

步骤6，浏览器获取响应报文并解析报文。

步骤7，关闭连接。

图 3.4　使用 IP 地址访问网页

3.3　报文

报文是指以一定格式组织起来的数据，分为请求报文和响应报文（3.2 节中已经提及）。其中，请求报文包括请求方法、请求的 URL、版本协议以及请求头信息。响应报文包括请求协议、响应状态码、响应头信息和响应内容。图 3.5 所示为某浏览器抓包获取的请求报文和响应报文信息。从图 3.5 中可以看到此次请求的方法是 GET；请求的 URL 为 http://www.********.com.cn/html/index.asp（该 URL 经过浏览器转换后的 IP 地址为 120.55.40.41，端口号为 80）；版本协议为 HTTP/1.1；请求头信息包括 Host、User-Agent 和 Accept 等。同时，从图 3.5 中也可以看出响应状态码为 200 OK；响应头信息包括 Cache-Control、Content-Type 和 Content-Encoding 等；具体的响应正文如图 3.6 所示。

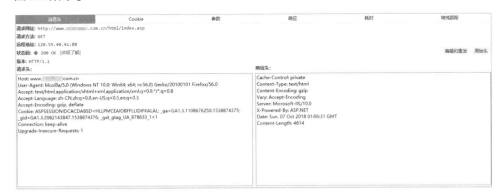

图 3.5　请求报文及响应报文

图 3.6　响应正文

3.4　HTTP 请求方法

在客户端向服务器发送请求时，需要确定使用的请求方法（也称为动作）。请求方法表明了对 URL 指定资源的操作方式，服务器会根据不同的请求方法进行不同的响应。在 HTTP/1.1 中，共定义了 8 种请求方法，具体如下所示。

- GET：发送请求获取服务器上的某特定资源。例如，HTML 文档、PDF 和图片等。GET 是相对常见的请求方法，但安全性较低。大多数网络爬虫，都采用 GET 获取网络资源。
- POST：向服务器提交数据，请求服务器进行处理。在通常情况下，表单提交操作需要使用 POST。
- HEAD：与 GET 类似，只会从服务器获取资源的头信息，不能获取响应内容。
- PUT：使用客户端向服务器传送的数据取代指定的内容。
- DELETE：请求服务器删除指定资源。
- CONNECT：在客户端配置代理的情况下，使用 CONNECT 建立客户端与服务器之间的联系。
- OPTIONS：询问服务器支持的请求方法，允许客户端查看服务器的性能。
- TRACE：对可能经过代理服务器传送到服务器上的报文进行追踪。

目前，这 8 种请求方法使用较多的是 GET 和 POST。另外，需要注意的是这些请求方法的名称是区分大小写的。

3.5　HTTP 状态码

HTTP 状态码由 3 位数字组成，描述了客户端向服务器请求过程中发生的状况。状态码的第一个数字描述了状况类型（成功、重定向等）。HTTP 状态码共有 5 种类型，如表 3.1 所示。

表 3.1　HTTP 状态码类型

数　　字	类　　型	描　　述
1XX	信息性状态码	服务器收到请求，需要请求者继续执行操作
2XX	响应成功状态码	客户端的请求成功并被服务器处理，返回响应内容
3XX	重定向状态码	客户端请求的 URL 被转移到新的 URL，需要进行附加操作以完成请求
4XX	客户端错误状态码	客户端请求的语法错误或网页不存在
5XX	服务器端错误状态码	服务器在处理请求时发生错误

下面介绍一些常见且具有代表性的状态码。

3.5.1 状态码 2XX

- 200：表示客户端发送的请求已被服务器正常处理。例如，在图 3.5 所示的状态码为 200 OK。一般在编写网络爬虫程序过程中，需要判断状态码是否为 200，如果为 200 则表示成功获取实体数据（如 HTML、JSON 等）。

```
if(StatusCode == 200){      //状态码200表示响应成功
    //获取实体内容
    //解析实体内容
}
```

- 204：表示客户端发送的请求已被服务器正常处理，但在返回的响应报文中不包含实体的主体部分。

3.5.2 状态码 3XX

- 301：永久性重定向，表示请求资源已被分配到新的 URL。
- 302：临时性重定向，表示请求资源已被临时分配到新的 URL，希望用户能使用新的 URL 访问，如用户登录某网站后会通过 302 跳转到登录后的主页，如图 3.7 所示。
- 303：临时重定向，告知客户端请求的资源存在另一个 URL，应使用 GET 方法定向获取请求资源。

图 3.7　状态码 302

3.5.3 状态码 4XX

- 403：表示客户端的请求被服务器拒绝，一般不说明拒绝的原因。使用网络爬虫短时间内获取大量数据时，会被服务器认定为攻击行为，服务器可能产生拒绝行为，同时本机的 IP 也会被封。
- 404：表示服务器无法找到 URL 对应的资源。通常服务器会返回一个实体，以便在客户端展示给用户，如图 3.8 所示。

图 3.8　状态码 404

3.5.4 状态码 5XX

- 500：表示服务器在执行请求时发生了错误，也有可能是 Web 应用存在 bug 或临时故障。
- 503：表示服务器暂时处于超负载或正在进行停机维护状态，无法处理请求。图 3.9 所示为请求某网页，出现状态码 503 的界面，其提示信息为"No server is available to handle this request"。

图 3.9　状态码 503

3.6　HTTP 信息头

HTTP 信息头，也称为头字段或首部，是构成 HTTP 报文的要素之一，具有传递额外重要信息的作用。HTTP 信息头通常包括 4 类：通用头（General Header）、请求头（Request Header）、响应头（Response Header）和实体头（Entity Header）。其中，请求头和响应头分别只在请求信息和响应信息中出现，而通用头和实体头在请求信息和响应信息中都可出现。只有在消息中包含实体数据时，实体头才会出现。

HTTP 信息头是由头字段名和字段值组成的，如下所示。

```
//请求头
Host: www.********.com.cn
User-Agent: Mozilla/5.0 (Windows NT 10.0; Win64; x64; rv:56.0)
Gecko/20100101 Firefox/56.0
Accept: text/html,application/xhtml+xml,application/xml;q=0.9,
*/*;q=0.8
//响应头
Server: Microsoft-IIS/10.0
Date: Tue, 09 Oct 2018 00:09:08 GMT
Content-Length: 4614
```

下面将分别介绍这 4 类信息头。

3.6.1 通用头

通用头既可以在请求信息中出现也可以在响应信息中出现，其提供了与报文相关的基本信息。表 3.2 列出了 HTTP 通用头的字段名及其功能。

表 3.2 HTTP 通用头的字段名及其功能

字 段 名	功 能
Cache-Control	请求和响应遵循的缓存机制
Connection	客户端和服务器指定与请求或响应连接有关的选项，例如是否需要持久连接
Date	创建 HTTP 报文的时间，即信息发送时间
Pragma	包含用来实现特定的指令，通常用 Pragma:no-cache
Trailer	表明以 chunked 编码传输的报文实体数据尾部存在的字段
Transfer-Encoding	规定了传输报文实体数据采用的编码方法
Upgrade	检测 HTTP 协议，允许服务器指定一种新的协议
Via	追踪客户端与服务器之间的请求报文和响应报文的传输路径（网管、代理服务器等）
Warning	告知用户与缓存相关的警告

- Cache-Control：指令可按请求和响应分类。请求消息中的缓存指令包括 no-cache、no-store、max-age、max-stale、min-fresh 和 only-if-cached 等，响应消息中的指令包括 public、private、no-cache、no-store、no-transform、must-revalidate、proxy-revalidate 和 max-age 等。表 3.3 所示为 Cache-Control 请求指令及各指令说明，表 3.4 所示为 Cache-Control 响应指令及各指令说明。

表 3.3　Cache-Control 请求指令

指　　令	说　　明
no-cache	告知服务器不直接使用缓存,目的是防止从缓存中返回过期的资源
no-store	提示请求或响应中包含机密信息,规定不缓存请求或响应中的任何内容
max-age	客户端希望接收存在时间不超多规定秒数的资源
max-stale	客户端希望接收存在时间超多规定秒数的资源
min-fresh	客户端希望接收还未超过指定秒数的缓冲资源
only-if-cached	客户端仅在服务器本地已缓存目标资源的情况下,才要求服务器返回资源

表 3.4　Cache-Control 响应指令

指　　令	说　　明
public	可向任意方提供响应缓存
private	仅向特定用户提供响应缓存
no-cache	不能直接使用缓存,要向服务器发起验证
no-store	提示请求或响应中包含机密信息,规定不缓存请求或响应中的任何内容
no-transform	不得对资源进行转换或转变
max-age	告知客户端,资源在规定的秒数内是最新的,无需向服务器发出新请求
must-revalidate	可缓存但必须由服务器发出验证请求,请求失败返回 504
proxy-revalidate	要求中间缓存服务器(如代理)对缓存的响应有效性再进行确认

下面为请求指令的一个案例。

```
Cache-Control: no-cache
Cache-Control: no-store
Cache-Control: max-age=0
Cache-Control: max-stale=3000
Cache-Control: min-fresh=60
Cache-Control: only-if-cached
```

下面为响应指令的一个案例。

```
Cache-Control: must-revalidate
Cache-Control: no-cache=Location
Cache-Control: no-store
Cache-Control: no-transform
Cache-Control: public
Cache-Control: private
Cache-Control: proxy-revalidate
//可以使用多个值
Cache-Control: private, s-maxage=0, max-age=0, must-revalidate
```

- Connection：主要指令有 Upgrade、keep-alive、close。Upgrade 用于检测协议是否可以使用更高版本连接控制不在转发给代理头字段。HTTP/1.1 版本的请求默认使用持久连接，当服务器想断开连接时，则 Connection 指定为 close。HTTP/1.1 之前的 HTTP 版本默认使用非持久连接，当客户端想要保持网络连接打开时，则 Connection 指定为 keep-alive。

```
Connection: Upgrade
Connection: keep-alive
Connection: close
```

- Date：用于创建 HTTP 报文的时间，如下所示。

```
Date: Tue, 09 Oct 2018 00:09:08 GMT
```

- Pragma：HTTP/1.1 之前版本中的通用头，仅作为向后兼容只支持 HTTP/1.0 协议的缓存服务器。规范定义的形式唯一，如下所示（效果与 HTTP/1.1 协议中的 Cache-Control: no-cache 相同）。

```
Pragma: no-cache
```

- Trailer：一个响应头，允许服务器在发送的报文主体后面添加额外的内容。Trailer 后面的字段值为报文主体后面所要追加的字段名称。如下案例所示。

```
HTTP/1.1 200 OK
Content-Type: text/html
Transfer-Encoding: chunked
Trailer: Expires
//HTML内容
0
Expires: Tue, 09 Oct 2018 00:09:08 GMT
```

在案例中，指定 Trailer 字段值为 Expires，在报文主体后面（0 分割）追加了字段 Expires，内容为 "Tue, 09 Oct 2018 00:09:08 GMT"。

- Transfer-Encoding：主要指令有 chunked、compress、deflate、gzip、identity，每种指令代表一种数据压缩算法。如下案例所示。

```
Transfer-Encoding: chunked
Transfer-Encoding: compress
Transfer-Encoding: deflate
Transfer-Encoding: gzip
Transfer-Encoding: identity
```

```
//可以使用多个值
Transfer-Encoding: gzip, chunked
```

- Upgrade：用于向服务器指定某种传输协议以便服务器进行转换，使用头字段 Upgrade 时，需要额外指定 Connection: Upgrade。如下案例所示。

```
Upgrade: HTTP/2.0, SHTTP/1.3, IRC/6.9, RTA/x11
```

- Via：由代理服务器添加，适用于正向和反向代理，在请求和响应头中均可出现。用法需要规定所使用的协议版本号、公共代理的 URL 及端口号、内部代理的名称或别名等。如访问 "https://developer.*******.org/zh-CN/" 便可看到 Via 通用头（某浏览器抓包分析），如下所示。

```
via: 1.17fdbb51d3898069adb076093148ebbc6.**********.net (**********)
```

- Warning：使用 Warning 时需要规定所使用的三位数字警告码（如 110、111、112、113 等）、添加到 Warning 首部的服务器或软件的名称或伪名称（当代理未知的时候可以用 "-" 代替）、描述错误信息的警告文本和日期时间（可选项）。如下案例所示。

```
Warning: 112 - "cache down" " Tue, 09 Oct 2018 00:09:08 GMT "
```

3.6.2 请求头

请求头是从客户端向服务器发送请求报文时所用的字段。服务器根据请求头信息，为客户端提供响应。在网络爬虫采集数据时，为了更好地模拟浏览器访问服务器，经常需要设置一些请求头信息，比如添加多个不同的 User-Agent。表 3.5 所示为部分 HTTP 请求头的字段名及其功能。

表 3.5 HTTP 请求头的字段名及其功能

字 段 名	功 能
Accept	指定客户端可以处理的数据类型
Accept-Charset	指定客户端可以接收的字符集
Accept-Encoding	指定浏览器能够进行解码的数据编码格式
Accept-Language	指定浏览器可接收的语言种类
Cookie	客户端发送请求时，会把保存在该请求域名下的所有 cookie 值一起发送给服务器
Host	指定请求的服务器的域名和端口号，不包括协议

续表

字 段 名	功 能
Origin	指定请求的服务器名称,即包括协议和域名
Referer	告知服务器请求的原始资源的 URL,即包括协议、域名、端口等信息
Upgrade-Insecure-Requests	向服务器发送一个信号,表示客户对加密和认证响应的偏好
User-Agent	发起请求的应用程序名称

- Accept：使用形式为 type/subtype，使用时一般一次指定多种数据类型。3.1 节给出了一些常见的数据类型，下面给出了 Accept 的几个示例。

```
Accept: */*
Accept: text/css,*/*;q=0.1
Accept: text/html,application/xhtml+xml,application/xml;
q=0.9,*/*;q=0.8
Accept:text/html,application/xhtml+xml,application/xml;q=0.9,im
age/webp,image/apng,*/*;q=0.8
```

在上面的几个示例中，*代表所有，每种数据类型之间用逗号分隔；q 表示给数据类型设置优先级，其取值范围是(0,1]，如果不设置默认为 1。

- Accept-Charset：使用形式与 Accept 类似，使用时一般一次指定多种字符集。下面给出了一个示例。

```
Accept-Charset: utf8, gbk; q=0.8
```

- Accept-Encoding：常用的数据编码方式有 gzip、deflate、br、sdch、compress 和 identity。下面给出了两个示例。

```
Accept-Encoding:gzip, deflate, br
Accept-Encoding:gzip, deflate, sdch, br
```

- Accept-Language：常用的语言种类有 zh-CN（简体中文）、zh（中文）、en-US（英语－美国）和 en（英语）。使用时，可以设置语言种类的优先级，下面给出了一个示例。

```
Accept-Language:zh-CN,zh;q=0.9,en;q=0.8
```

- Cookie：语法为 Cookie: name=value，包含多个 name 和 value 值的，中间使用";"字符分割，下面给出了一个示例。

```
Cookie: WMF-Last-Access=10-Oct-2018; WMF-Last-Access-Global=10-
Oct-2018; GeoIP=CN:AH:Hefei:31.86:117.28:v4
```

- Host：所有的 HTTP 请求必须携带 Host 头。如服务器未设定主机，则 Host 为空值。下面分别给出了请求某网站页面以及某大学新闻页面的对应的 Host。

```
Host: www.*****.com
Host:news.****.edu.cn
```

- Origin：一般用于 CORS（Cross-origin resource sharing）跨域请求和 POST 请求中，下面给出了请求谷歌学术得到的 Origin。

```
Origin:https://scholar.******.com
```

- Referer：浏览器向服务器发送请求主要有两种方式，即用户在浏览器中直接输入 URL；用户通过单击某网页上的超链接访问 URL。通常情况下，使用第一种方式浏览器不会发送 Referer 信息给服务器，而使用第二种方式则会发送 Referer 信息给服务器。因此，Referer 常被网站管理人员用来追踪网站的访问者，并查看其是如何导航进入网站的。如在谷歌学术网站中输入关键词"Lda"并单击搜索按钮，如图 3.10 所示，通过浏览器抓包便可以发现请求头中存在 Referer，而直接在浏览器输入网址便不会在请求头中找到 Referer。在网络爬虫中，经常在请求的 headers 中伪造 Referer 来绕过网站的一些防爬措施。

```
Referer:https://scholar.******.com/scholar?hl=zh-CN&as_sdt=0%2C
5&q=Lda&btnG=
```

- Upgrade-Insecure-Requests：用于升级不安全请求，指示浏览器在进行 URL 请求之前升级不全的网站，如"http://blog.****.net"会升级成"https://blog.****.net"。下面是 Upgrade-Insecure-Requests 头的用法。

```
Upgrade-Insecure-Requests: 1
```

图 3.10　谷歌学术网站

- User-Agent：用于标识请求浏览器的身份。目前，很多网站为了防止自身数据被网络爬虫采集，都会对请求头进行校验，其中重要的校验对象便是 User-Agent。而网络爬虫为了更好地采集数据，通常需要添加 User-Agent 库，每次请求 URL 时，可以从 User-Agent 库中随机挑选一个使用，进而达到伪造浏览器安全访问服务器资源的目的。程序 3-1 演示了在 Java 中创建 User-Agent 库的一种方式。

程序 3-1

```
//阿里巴巴开发的集合操作工具
List<String> userAgent = Lists.newArrayList(
            "Mozilla/5.0 (Macintosh; U; Intel Mac OS X 10_6_8; en-us) AppleWebKit/534.50 (KHTML, like Gecko) Version/5.1 Safari/534.50",
            "Mozilla/5.0 (Windows; U; Windows NT 6.1; en-us) AppleWebKit/534.50 (KHTML, like Gecko) Version/5.1 Safari/534.50",
            "Mozilla/5.0 (compatible; MSIE 9.0; Windows NT 6.1; Trident/5.0)",
            "Mozilla/4.0 (compatible; MSIE 7.0; Windows NT 6.0)",
            "Mozilla/4.0 (compatible; MSIE 6.0; Windows NT 5.1)",
            "Mozilla/5.0 (Macintosh; Intel Mac OS X 10.6; rv:2.0.1) Gecko/20100101 Firefox/4.0.1",
            "Mozilla/5.0 (Windows NT 6.1; rv:2.0.1) Gecko/20100101 Firefox/4.0.1",
            "Opera/9.80 (Windows NT 6.1; U; en) Presto/2.8.131 Version/11.11",
            "Mozilla/5.0 (Macintosh; Intel Mac OS X 10_7_0) AppleWebKit/535.11 (KHTML, like Gecko) Chrome/17.0.963.56 Safari/535.11",
            "Mozilla/4.0 (compatible; MSIE 7.0; Windows NT 5.1; Maxthon 2.0)"
    );
```

3.6.3 响应头

响应头是从服务器端向客户端发送响应报文时所用的字段。表 3.6 所示为部分 HTTP 响应头的字段名及其功能。

表 3.6 HTTP 响应头的字段名及其功能

字 段 名	功 能
Accept-Ranges	指定服务器对资源请求的可接受范围类型，字段的值定义了范围类型的单位
Age	服务器产生响应经过的时间，单位为秒，为非负整数，主要用于缓存
Set-Cookie	用来由服务器端向客户端发送 cookie
Server	指明服务器软件以及版本号
Vary	告知代理是使用缓存响应还是从源服务器中重新请求资源

下面给出了如表 3.6 所示响应头的示例。

```
Accept-Ranges: bytes
Age: 1296
Set-Cookie: BDRCVFR[Fc9oatPmwxn]=mk3SLVN4HKm; path=/; domain
=.baidu.com
Vary: Accept-Encoding
Server: BWS/1.1
```

3.6.4　实体头

请求报文和响应报文中经常包含一些实体数据，如浏览器采用 POST 提交的表单数据、服务器返回给浏览器的网页数据等。实体头提供了大量的有关实体数据的信息，包括实体数据的类型、长度和压缩方法等。表 3.7 所示为部分 HTTP 实体头的字段名及其功能。

表 3.7　HTTP 实体头的字段名及其功能

字　段　名	功　　能
Allow	列出资源所支持的 HTTP 方法集合
Content-Encoding	告知客户端服务器对实体数据的编码方式
Content-Language	告知客户端实体数据使用的语言类型
Content-Length	实体数据的长度
Content-Location	实体数据的资源位置
Content-Range	当前传输的实体数据在整个资源中的字节范围
Content-Type	实体数据的类型
Expires	实体数据的有效期
Last-Modified	实体数据上次被修改的日期及时间

- Allow：告知客户端服务器能够支持的请求方法，当服务器接收不到支持的请求方法时，会以 405 Method Not Allowed 作为响应返回。Allow 示例如下所示。

```
Allow: GET, POST, HEAD
```

- Content-Encoding：常用的数据编码方式有 gzip、deflate、br、sdch、compress 和 identity。Content-Encoding 示例如下。

```
Content-Encoding: gzip
Content-Encoding: compress
Content-Encoding: br
```

- Content-Language：常用的语言种类有 zh-CN（简体中文）、zh（中文）、en-US（英语—美国）和 en（英语）。Content-Language 示例如下。

```
Content-Language: en
Content-Language: zh-CN
```

- Content-Length：单位是字节。注意当对实体数据进行编码传输时，不适用 Content-Length 头。Content-Length 示例如下。

```
Content-Length: 11410
```

- Content-Location：当客户端请求的资源在服务器有多个地址时，服务器可以通过 Content-Location 字段告知客户端其他可选地址。Content-Location 示例如下。

```
Content-Location: /index.html
```

- Content-Range：主要用于针对范围请求，即告知客户端返回响应实体的哪个部分符合范围请求，字段值以字节为单位。Content-Range 示例如下。

```
Content-Range: bytes 800-5000/67589
```

- Content-Type：使用形式与请求头 Accept 类似，字段值用 type/subtype 表示，另外也可以在 Content-Type 中对实体数据的字符集进行设置。Content-Type 示例如下。

```
Content-Type: text/html; charset=UTF-8
```

- Expires：如果在响应头 Cache-Control 设置了 max-age 指令，则 Expires 可以被忽略。Expires 示例如下。

```
Expires: Thu, 11 Oct 2018 01:23:51 GMT
```

- Last-Modified：指 URL 对应资源的最后一次修改时间。Last-Modified 示例如下（请求 wikipedia 可以看到该实体头）。

```
Last-Modified: Wed, 03 Oct 2018 13:57:31 GMT
```

3.7 HTTP 响应正文

HTTP 响应正文（或 HTTP 响应实体主体）指服务器返回的一定格式的数据。在 3.5 节中，介绍了 Accept 请求头与 Content-Type 实体头。其中，Accept 表示客户端可以接收响应正文的数据类型，字段值可以有多个，而 Content-Type 主要表示服务器返回响应正文的数据类型，仅有一个字段值。在本节中将主要介绍网络爬虫中经常解析的几种数据类型：HTML、XML 和 JSON。

3.7.1 HTML

HTML（HyperText Markup Language），即超文本标记语言，是一种用于创建网页的标准标记语言。HTML 描述并定义了网页的内容和布局，其中，网页内容包括文本、图片和表格数据等，网页布局可以使用<div>元素和<table>元素等。服务器返回的响应正文多以 HTML 的形式呈现。下面以一个简单的实例展示 HTML 的结构，如程序 3-2 所示。

程序 3-2

```
<!DOCTYPE html>
<html>
<head>
<meta charset="UTF-8">
<title>测试网页</title>
</head>
<body>
  <div id="maincontent">
    <div id="********">
        <h1>浏览器脚本教程</h1>
        <p>
            <strong>从左侧的菜单选择你需要的教程！</strong>
        </p>
    </div>
    <div class="item" id="i1" >
        <h2>JavaScript</h2>
        <p>JavaScript 是世界上最流行的脚本语言。</p>
        <p>
            <a href="/js/index.asp" title="JavaScript教程">JavaScript 教程</a>。
        </p>
    </div>
    <div class="item" id="i2">
        <h2>HTML DOM</h2>
        <p>HTML DOM 定义了访问和操作 HTML 文档的标准方法。</p>
        <p>
            <a href="/htmldom/index.asp" title="HTML DOM 教程">
```

开始学习 HTML DOM</a>！
            </p>
        </div>
        <div class="card-handle" >
            <img id="dimg_16" src="https://image.*****.com/wp-content/uploads/2017/03/uisdc-pic-20170327-3.jpg" width="160">
        </div>
    </div>
</body>
</html>

通过程序 3-2 可以发现 HTML 为相互嵌套且布局规律的树型结构，如图 3.11 所示。下面对程序 3-2 中的内容，进行简要说明。

- 文档类型声明：<!DOCTYPE html>声明为 HTML5 文档。
- <html>元素：HTML 页面的根元素。
- <head>元素：包含了文档的元（meta）数据，定义网页的编码形式如 utf-8。
- <title>元素：网页文档的标题。
- <body>元素：包含了可见的页面内容。
- <div>元素：用于创建布局。
- <h1>元素：定义标题。
- <h2>元素：定义标题。
- <p>元素：定义段落。
- <a>元素：用于文本超链接。
- <img>元素：图像标签。

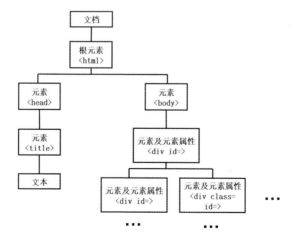

图 3.11　HTML 树型结构表示

一般情况下，元素由一对标签表示，即开始标签与结束标签，如上面实例中的<p>与</p>。元素可以设置属性，属性在开始标签中描述，如<div id="maincontent">中 id 为属性，值为"maincontent"。表 3.8 所示为 HTML 元素常见的几个属性。

表 3.8 HTML 元素中常见的几个属性

属性	值	描述
class	classname	规定元素的类名
id	id	规定元素的唯一 id
style	style_definition	规定元素的行内样式
title	text	规定元素的额外信息

图 3.12 程序 3-2 中的 HTML 实例对应的页面

图 3.12 所示为程序 3-2 中 HTML 实例对应的页面。HTML 文件中通常包括很多内容，可以利用 HTML 的结构解析得到符合自身需求的数据，如文本、超链接和网页标题等。关于如何使用 Java 程序解析 HTML 文件，将在第 5 章中详细介绍。

3.7.2 XML

XML（Extensible Markup Language，简称 XML），即可扩展标记语言，用于传输和存储数据。XML 也是树型结构，一般由 XML 声明、元素、元素属性和文本等信息组成，和 HTML 类似，但两者不可互相替代。除了 HTML，XML 也是网站经常返回的一种数据类型。另外，XML 也经常用作程序的配置文件，如 Maven 工程中的 pom.xml、Spring MVC 架构中的 web.xml 等。程序 3-3 所示为 XML 的一个实例。

程序 3-3

```xml
<?xml version="1.0" encoding="UTF-8"?>
<books>
 <book category="java">
    <title>Java编程思想</title>
    <author> Bruce Eckel</author>
 </book>
 <book category="C++">
    <title>C++ Primer Plus</title>
    <author>Stephen Prata</author>
 </book>
</books>
```

在程序 3-3 中，第一行为 XML 声明，描述了 XML 的标准和编码。XML 中的元素也是由开始标签与结束标签表示的，如<books>与</books>、<title>与</title>等。其中,标签<book>中的category 表示属性。图 3.13 所示为上述 XML 实例的树型结构表示。

从结构上看，HTML 和 XML 具有一定的相似性。所以，在解析两种数据时，可以使用一些相同的工具（如 Jsoup），具体将在第 5 章介绍。

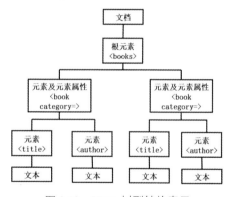

图 3.13　XML 树型结构表示

3.7.3　JSON

JSON（JavaScript Object Notation），即 JavaScript 对象表示法，是一种基于纯文本的轻量级数据表示方法。JSON 以 key:value 键/值对的方式记录数据，简单易读。目前，JSON 格式的数据在 Web 开发中越来越受重视，其通常用于网站上数据的表示和传输。特别是在使用 AJAX 开发项目的过程中，经常利用 JSON 将数据返回到前端。

JSON 的基本语法较为简单，如下所示。

数据在键/值对中，如 "id": "01"
键/值对使用逗号分隔
花括号保存对象
方括号保存数组

程序 3-4 展示了一个简单的 JSON 字符串。

程序 3-4

```
{
    "books": [
        {
            "id":"01",
            "language": "Java",
            "edition": "third",
            "author": "Herbert Schildt"
        },
        {
            "id":"07",
            "language": "C++",
            "edition": "second",
            "author": "E.Balagurusamy"
        }]
}
```

在网络爬虫实战中，经常遇到网页返回的数据类型为 JSON，如某电商网站的商品评价信息（见图 3.14）。

图 3.14　某电商网站的商品评价信息

另外，JSON 类型的数据也常通过<script>元素嵌入 HTML，如淘宝搜索页面就采用这种方式展示用户搜索关键词对应的产品。为便于理解，程序 3-5 给出了一个简单的实例，该程序对应的页面如图 3.15。使用网络爬虫解析这类页面时，首先需要抽取<script>元素中的 JSON 字符串，之后再使用 JSON 解析工具（如 Fastjson）解析 JSON 中的具体字段。

程序 3-5

```
<!DOCTYPE html>
<html>
<head>
<meta charset="utf-8">
<title>Json</title>
</head>
<body>
 <h2>Java书籍</h2>
 <p>
     书籍名称：<span id="bookname"></span><br /> 书籍地址：<span id="bookurl"></span><br />
     书籍作者：<span id="bookauthor"></span><br />
 </p>
 <script>
     var JSONObject = {
         "bookname" : "Java编程思想（第4版）",
         "bookurl" : "https://item.**.com/10058164.html",
         "bookauthor" : " Bruce Eckel! "
     };
     document.getElementById("bookname").innerHTML = JSONObject.bookname
     document.getElementById("bookurl").innerHTML = JSONObject.bookurl
     document.getElementById("bookauthor").innerHTML = JSONObject.bookauthor
 </script>
</body>
</html>
```

图 3.15　程序 3-5 对应的页面

3.8　网络抓包

3.8.1　简介

抓包（Packet Capture）是指对网络传输中发送与接收的数据包进行截获、重发、编辑和转存等操作。

在开发网络爬虫时，给定 URL，开发者必须清楚客户端是如何向服务器发送请求的，以及客户端发出请求后服务器返回的数据是什么。只有了解这些内容，开发者才能在程序中拼接 URL，针对服务返回的数据类型制定具体的解析策略。因此，网络抓包是实现网络爬虫必不可少的技能之一，也是网络爬虫开发的起点。

例如，采集某网购商品的评价数据时，只需要在浏览器中输入该商品的 URL，单击商品评价板块，便可以看到商品的评价信息（URL 没有发生变化），如图 3.16 所示。

图 3.16　某网购商品评价页面

在编写网络爬虫时，发现直接请求该 URL 获取的 HTML 中，却不包含该商品的评论信息。原因是在单击商品评价板块时，客户端向服务器发送了其他请求，而该请求的 URL 并不是 https://item.**.com/6773559.html，而是下面形式的 URL。

存储评价信息的 JSON 地址为 https://sclub.**.com/comment/…callback=fetchJSON_comment98vv4854&productId=6773559…

实际上，基于网络抓包很容易捕获浏览器每一步操作对应的请求信息以及响应信息。图 3.14 所示为单击商品评价板块后，服务器响应内容的结构。

3.8.2 使用情境

对很多初学者而言，由于不了解网络抓包，而直接去请求某个 URL（如上节中介绍的案例），可能导致无法获取自己想要的数据。在此，对网络抓包的使用情境做如下说明。

（1）直接请求 URL 能够获得数据时，可以不用进行网络抓包分析。但当使用程序直接请求 URL 无法找到网页中的数据时（多为 JSON），则需要采取网络抓包措施，获取数据对应的真实请求 URL。

（2）执行表单请求时，需要采取网络抓包措施，如需要模拟登录网站才能获取的数据。基于网络抓包，可以捕获表单请求提交的详细参数（如用户名、密码和 Public Key 等）。

（3）在将捕获请求信息添加到程序中，如利用网络抓包获取一些浏览器请求网页的头信息和请求方法等。

3.8.3 浏览器实现网络抓包

网络抓包分析的内容主要是报文信息，即 HTTP 请求方法、信息头、响应状态码、响应正文内容和表单的提交参数等。

多数浏览器，都具有 HTTP 抓包的功能。下面使用火狐浏览器和谷歌浏览器进行操作。

（1）火狐浏览器抓包

本案例使用火狐浏览器（版本为 56.0 64 位）对某搜索页面进行抓包分析，分析的 URL 与搜索关键词如下所示。

URL：http://search.****.com.cn/。

搜索的关键词：金融。

具体流程如下所示。

步骤 1，在浏览器中，右击"查看元素"按钮，在弹出框中单击"网络"按钮，如图 3.17 所示。

图 3.17　火狐浏览器抓包步骤 1

步骤 2，在该网页的搜索框中输入关键字，如"金融"，并单击"搜索新闻"按钮。可以看到页面进行了跳转，并产生了一系列的请求，如图 3.18 所示。

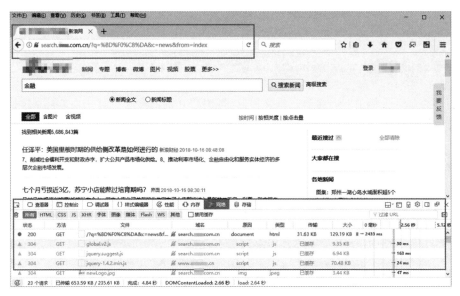

图 3.18　火狐浏览器抓包步骤 2

步骤 3，找到个人需要采集的数据（如新闻标题、新闻地址、新闻来源等）对应的具体请求，并单击。之后，可以看到关键词"金融"对应请求的真实 URL、请求方法、响应状态码、HTTP 协议版本和信息头，如图 3.19 所示。

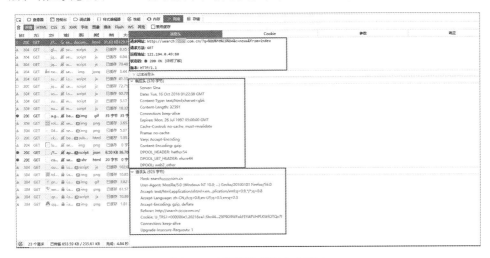

图 3.19 火狐浏览器抓包步骤 3

由图 3.19 可以看出，关键词"金融"对应的 GBK 编码为%BD%F0%C8%DA，向后台发送请求的真实 URL 为 http://search.****.com.cn/?q=%BD%F0%C8%DA&c=news&from=index

请求地址中的 q=%BD%F0%C8%DA、c=news 和 from=index 为参数部分。在 Java 中，java.net 包提供了 URL 中文的编码的处理方法，如程序 3-6 所示。

程序 3-6

```java
package com.qian.urlEncoded;

import java.io.UnsupportedEncodingException;
import java.net.URLEncoder;
public class UrlEncoded {
    public static void main(String[] args) {
        String keyword = "金融";                    //需要编码的关键词
        try {
            //输入编码方式为GBK
            String keywordEncoded = URLEncoder.encode(keyword, "gbk");
            System.out.println(keywordEncoded);    //输出结果%BD%F0%C8%DA
        } catch (UnsupportedEncodingException e) {
            e.printStackTrace();
```

		}
	}
}

步骤4，单击"响应"，可以看到详细的响应正文，如图3.20所示。

图 3.20　火狐浏览器抓包步骤4

（2）谷歌浏览器抓包

网络抓包能够分析模拟登录时，可以到具体的请求 URL 以及表单提交参数。下面，利用谷歌浏览器（版本为 63.0 64 位），分析某网站登录页面，如图 3.21 所示。

图 3.21　某网站登录页面

具体操作流程如下所示。

步骤1，在谷歌浏览器中，右击"检查"按钮，在弹出框中单击"Network"，勾选"Preserve log（保存日志）"，如图 3-22 所示。刷新页面，并在账户登录页面输入用户名及密码。

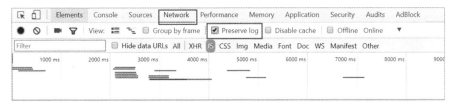

图 3.22　谷歌浏览器抓包

步骤 2，单击"登录"按钮，执行网络抓包。在"Name"栏中找到与登录提交参数有关的请求。图 3.23 为网络抓包得到的登录过程中表单提交的内容，共有 12 项参数：uuid、eid、fp、_t、loginType、loginname、nloginpwd、authcode、pubKey、sa_token、seqSid 和 useSlideAuthCode。同时，在"General"可以看到具体的请求 URL。

```
Request URL: https://passport.**.com/uc/loginService?uuid=82d76f63
-c0e1-4efe-8c48-35d6ab8c0891&&r=0.9900400911798082&version=2015
```

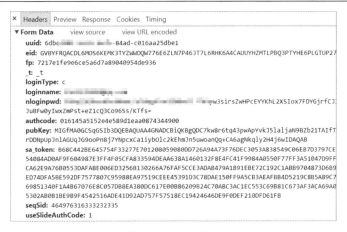

图 3.23　登录过程中表单提交的内容

步骤 3，步骤 2 中的表单提交的参数不仅包含用户名及密码，还包含一些其他参数。这些参数可以在 https://passport.**.com/new/login.aspx 对应的 HTML 中找到，如图 3.24 所示。同时，在图 3.24 中，可以发现客户端向服务器提交的密码是经过加密的，加密算法为 RSA，该算法依赖于网页中的公钥（Public key）。

在网络爬虫模拟登录过程中，需要提交这些参数。所以，首先需要获取 https://passport.**.com/new/login.aspx 页面对应的具体参数值。接着，基于 pubKey 参数的值，使用 RSA 算法，获取加密后的密码。其次，拼接登录的 Request URL（这里只需使用 uuid）。最后，提交所有参数及参数值便可达到模拟登录该网站的目的。

```
▼<form id="formlogin" method="post" onsubmit="return false;">
    <input type="hidden" id="sa_token" name="sa_token" value=
    "B68C442BE645754F33277E70120805908ODD726A94A73F76DEC3053A838549C06EB7D3797CE1C5BBE7C2B2EF9CA7D467C3C76FF0A28885EE64B432
    348C6987D2A54084AD0AF9F604987E3FF4F05CFA833594DEAA638A1460132F8E4FC41F9984A0550F77FF3A51047D9FFA6937B2323ADE6CDB3A98776
    0D104BE5A33FD482D219E65930F424CA62E9A76B0553DAFABE006ED3256B130266A76FAF5CCE3ADAB479A1B91EBE72C192C1ABB9704873D6B9E993E
    59961E92ACE3581837CC7E60A9C7DF01386E5A8CBD03AEEC4ED74DFA5BE592DF7577807C95988EA97519CEEE45391D3C78DAE150FF9A5CB3AEAFBB4
    89C735FF957448D469385C09BBD19D83A975242FCF0AEB8044F47CA93E3F0956D7C569851340F1A4B67076E8C057DB8EA380DC617E00B86209B24C7
    553C69B81C673AF3ACA69A033B730BEAA3E5AE5EA9C59235C87E5D8E9D48D70DDB3283B3589B61E10D91B40546C71B483B4215633CFCE77CD0A9CAE
    11F03658A67F34F33788EB08B69">
    <input type="hidden" id="uuid" name="uuid" value="a3b43426-90a8-4b23-86b6-19c1cfd449b1">
    <input type="hidden" name="eid" id="eid" value=
    "GVBYFRQACDL6MOS6KEMK3TYZWWOQW776E6ZLN7P46JT7L6RHK6A4CAUUYHZMTLPBQ3PTYHE6PLGTUP27PRXV6YAXZA" class="hide">
    <input type="hidden" name="fp" id="sessionId" value="7217e1fe9e6ce5a6d7a89040954de936" class="hide">
    <input type="hidden" name="_t" id="token" value="_t" class="hide">
    <input type="hidden" name="loginType" id="loginType" value="c" class="hide">
    <input type="hidden" name="main_flag" id="main_flag" value="main_flag" class="hide">
    <input type="hidden" name="pubKey" id="pubKey" value=
    "MIGfMA0GCSqGSIb3DQEBAQUAA4GNADCBiQKBgQDC7kw8r6tq43pwApYvkJ5laljaN9BZb21TAIfT/
    vexbobzH7Q8SUdP5uDPXEBKzOjx2L28y7Xs1d9v3tdPfKI2LR7PAzWBmDMn8riHrDDNpUpJnlAGUqJG9ooPn8j7YNpcxCa1iybOlc2kEhmJn5uwoanQq+CA
    4j6wIDAQAB" class="hide">
    <input type="hidden" id="slideAppId" value="1604ebb2287" class="hide">
    <input type="hidden" name="useSlideAuthCode" id="useSlideAuthCode" value="1" class="hide">
  ▶<div class="item item-fore1">…</div>
```

图 3.24　HTML 中的其他参数

3.8.4　其他网络抓包工具推荐

（1）Fiddler：常用的网络抓包工具之一，能够记录客户端和服务器之间的所有 HTTP/HTTPS 请求，可针对特定的 HTTP/HTTPS 请求，分析请求数据、设置断点、调试 Web 应用和修改请求的数据，甚至可以修改服务器返回的数据。

（2）Wireshark：一款非常强大的网络抓包工具，可以截取各种网络封包，显示网络封包的详细信息。但在处理 HTTP/HTTPS 时，使用基本的浏览器抓包或 Fiddler 抓包就够了，如果处理 ARP/TCP/UDP/CIMP 时，可以使用 Wireshark。

（3）Postman：Google 开发的一款功能强大的用于网页调试与发送网页 HTTP/HTTPS 请求的工具（既有谷歌插件又有桌面软件）。在 Postman 中输入浏览器抓包的内容，可以实现模拟请求。

3.9　本章小结

本章主要介绍了 HTTP 有关的内容，掌握这些内容，将有利于网络爬虫项目的开发。例如，利用 POST 请求方法提交参数，以达到模拟登录网站的目的；通过 HTTP 状态码判断请求 URL 成功与否；通过随机切换请求头，以防止网络爬虫被服务器判断为机器人；根据不同的响应正文，使用不同的解析工具等。

第 4 章

网页内容获取

4.1 Jsoup 的使用

Jsoup 是一款基于 Java 语言的开源项目,主要用于请求 URL 获取网页内容、解析 HTML 和 XML 文档。使用 Jsoup 可以非常轻松地构建一些轻量级网络爬虫。在本节中,主要使用 Jsoup 获取网页内容,关于 Jsoup 的解析功能将在第 5 章进行介绍。

4.1.1 jar 包的下载

首先,在 Eclipse 中创建 Maven 工程,并在 Maven 工程的 pom.xml 文件中添加 Jsoup 对应的 dependency。

```
<!-- https://*************.com/artifact/org.jsoup/jsoup -->
<dependency>
    <groupId>org.jsoup</groupId>
    <artifactId>jsoup</artifactId>
    <version>1.11.3</version>
</dependency>
```

基于 pom.xml 文件的配置信息,可以下载 Jsoup 的 jar 包(版本是 1.11.3)。在 Jsoup 项目中,共包含 7 个 package,如图 4.1 所示。

图 4.1　Jsoup 项目中的 7 个 package

4.1.2 请求URL

org.jsoup.Jsoup 类可以用来处理连接操作。在 org.jsoup.Jsoup 类中提供了 connect(String url)方法来创建一个新连接，该方法的实现依赖于 Java 网络通信包 java.net。在创建连接之后，可通过具体请求方法（如 GET 和 POST）获取 URL 对应的 HTML 文件。

例如，采集某页面中的文本内容。首先，利用谷歌浏览器进行抓包，分析请求的URL、请求的方法、请求头、响应头等信息，如图 4.2 所示。

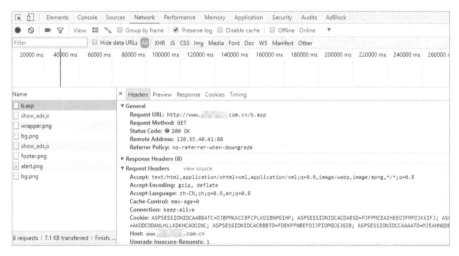

图4.2 谷歌浏览器抓包分析

程序 4-1 所示为 Jsoup 请求该网页的一种方式，这里使用 get()方法执行请求，该方法返回一个 Document 类型的对象，使用 Element 类中的 html()方法可以将 Document 类型的对象转化为 String 类型。图 4.3 所示为程序 4-1 的运行结果。

程序 4-1

```java
import java.io.IOException;
import org.jsoup.Connection;
import org.jsoup.Jsoup;
import org.jsoup.nodes.Document;
public class JsoupConnectURL1{
  public static void main(String[] args) throws IOException {
    //创建连接
    Connection connect = Jsoup.connect("http://www.********.com.cn/b.asp");
    //请求网页，也可以使用POST请求
```

```
        Document document = connect.get();
        //输出HTML
        System.out.println(document.html());
    }
}
```

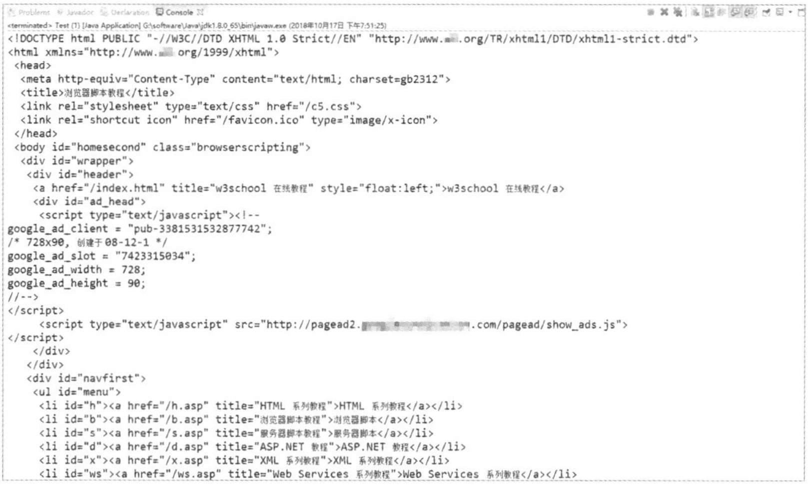

```
<terminated> Test (1) [Java Application] G:\software\Java\jdk1.8.0_65\bin\javaw.exe (2018年10月17日 下午7:51:25)
<!DOCTYPE html PUBLIC "-//W3C//DTD XHTML 1.0 Strict//EN" "http://www..org/TR/xhtml1/DTD/xhtml1-strict.dtd">
<html xmlns="http://www. org/1999/xhtml">
 <head>
  <meta http-equiv="Content-Type" content="text/html; charset=gb2312">
  <title>浏览器脚本教程</title>
  <link rel="stylesheet" type="text/css" href="/c5.css">
  <link rel="shortcut icon" href="/favicon.ico" type="image/x-icon">
 </head>
 <body id="homesecond" class="browserscripting">
  <div id="wrapper">
   <div id="header">
    <a href="/index.html" title="w3school 在线教程" style="float:left;">w3school 在线教程</a>
    <div id="ad_head">
     <script type="text/javascript"><!--
google_ad_client = "pub-3381531532877742";
/* 728x90, 创建于 08-12-1 */
google_ad_slot = "7423315034";
google_ad_width = 728;
google_ad_height = 90;
//-->
</script>
     <script type="text/javascript" src="http://pagead2..com/pagead/show_ads.js">
</script>
    </div>
   </div>
   <div id="navfirst">
    <ul id="menu">
     <li id="h"><a href="/h.asp" title="HTML 系列教程">HTML 系列教程</a></li>
     <li id="b"><a href="/b.asp" title="浏览器脚本教程">浏览器脚本</a></li>
     <li id="s"><a href="/s.asp" title="服务器脚本教程">服务器脚本</a></li>
     <li id="d"><a href="/d.asp" title="ASP.NET 教程">ASP.NET 教程</a></li>
     <li id="x"><a href="/x.asp" title="XML 系列教程">XML 系列教程</a></li>
     <li id="ws"><a href="/ws.asp" title="Web Services 系列教程">Web Services 系列教程</a></li>
```

图 4.3　程序 4-1 的运行结果

另外，也可以利用 Jsoup 先获取响应 Response，再获取 HTML 内容，如程序 4-2 所示。在获取响应时，method(Method method)方法中的参数为具体的 HTTP 请求方法，这里共有 8 种，如图 4.4 所示。在程序 4-2 中，设置的请求方法为 GET。程序的输出结果如图 4-5 所示，具体包括请求的 URL、响应状态码、响应数据类型、响应状态信息以及 HTML 文件。如果将程序 4-2 中的请求方法设置成 HEAD，重新运行程序，则会发现无法输出 HTML 文件，但能获取响应头信息。

程序 4-2

```
import java.io.IOException;
import java.net.URL;
import org.jsoup.Connection.Method;
import org.jsoup.Connection.Response;
import org.jsoup.Jsoup;
import org.jsoup.nodes.Document;
public class JsoupConnectURL2 {
  public static void main(String[] args) throws IOException {
      //获取响应
      Response response = Jsoup.connect
("http://www.********.com.cn/b.asp")
```

```
                .method(Method.GET).execute();
    URL url = response.url();                    //查看请求的URL
    System.out.println("请求的URL为:" + url);
    int statusCode = response.statusCode();   //获取响应状态码
    System.out.println("响应状态码为:" + statusCode);
    String contentType = response.contentType();
    //获取响应数据类型
    System.out.println("响应类型为:" + contentType);
    //响应信息 200-OK
    String statusMessage = response.statusMessage();
    System.out.println("响应信息为:" + statusMessage);
    //判断响应状态码是否为200
    if (statusCode == 200) {
            //通过这种方式可以获得响应的HTML文件
        String html = new String(response.bodyAsBytes(),"gbk");
        //获取HTML内容,但对应的是Document类型
        Document document = response.parse();
        //这里HTML和Document数据是一样的,但Document是经过格式化的
        System.out.println(html);
    }
  }
 }
```

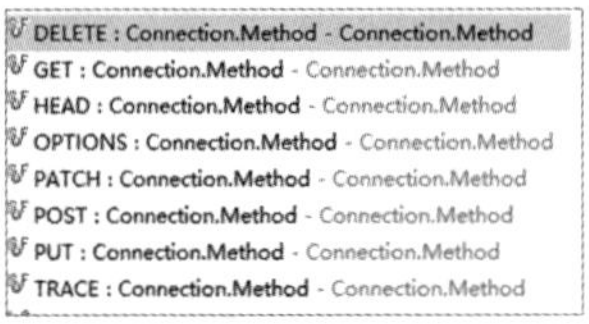

图 4.4 Method 中的 HTTP 请求方法

图 4.5 程序 4-2 的输出结果

4.1.3 设置头信息

在网络爬虫中，经常需要设置一些头信息。设置头信息的作用是伪装网络爬虫，使得网络爬虫请求网页更像浏览器访问网页，进而降低了网络爬虫被网站封锁的风险。Jsoup 中提供了两种设置头信息的方法，如下所示。

```
/**
 * Set a request header.
 * @param name header name
 * @param value header value
 * @return this Connection, for chaining
 * @see org.jsoup.Connection.Request#headers()
 */
Connection header(String name, String value);
/**
 * Adds each of the supplied headers to the request.
 * @param headers map of headers name {@literal ->} value pairs
 * @return this Connection, for chaining
 * @see org.jsoup.Connection.Request#headers()
 */
Connection headers(Map<String,String> headers);
```

第一种方法每次只可以设置一个请求头，如果要设置多个请求头，需要多次调用此方法；第二种方法可以添加多个请求头至 Map 集合。在程序 4-3 中，设置了一个请求头 User-Agent。在程序 4-4 中，设置了多个请求头，这些请求头来源于图 4.2 所示的网络抓包内容。

程序 4-3

```
import java.io.IOException;
import org.jsoup.Connection;
import org.jsoup.Jsoup;
import org.jsoup.nodes.Document;
public class JsoupConnectHeader {
  public static void main(String[] args) throws IOException {
      Connection connect = Jsoup.connect("http://www.********.com.cn/b.asp");
      //设置一个请求头
      Connection conheader = connect.header("User-Agent","Mozilla/5.0 (Windows NT 10.0; Win64; x64) AppleWebKit/537.36 (KHTML, like Gecko) Chrome/63.0.3239.108 Safari/537.36");
      Document document = conheader.get();
```

```java
        System.out.println(document);
    }
}
```

程序 4-4

```java
import java.io.IOException;
import java.util.HashMap;
import java.util.Map;
import org.jsoup.Connection;
import org.jsoup.Jsoup;
import org.jsoup.nodes.Document;
public class JsoupConnectHeaderMap {
  public static void main(String[] args) throws IOException {
        Connection connect = Jsoup.connect("http://www.********.com.cn/b.asp");
        //设置多个请求头。头信息保存到Map集合中
        Map<String, String> header = new HashMap<String, String>();
        header.put("Host", "www.********.com.cn");
        header.put("User-Agent", " Mozilla/5.0 (Windows NT 10.0; Win64; x64) AppleWebKit/537.36 (KHTML, like Gecko) Chrome/63.0.3239.108 Safari/537.36");
        header.put("Accept", "text/html,application/xhtml+xml,application/xml;q=0.9,image/webp,image/apng,*/*;q=0.8");
        header.put("Accept-Language", "zh-cn,zh;q=0.5");
        header.put("Accept-Encoding", "gzip, deflate");
        header.put("Cache-Control", "max-age=0");
        header.put("Connection", "keep-alive");
        Connection conheader = connect.headers(header);
        Document document = conheader.get();
        System.out.println(document);
    }
}
```

在网络爬虫实战中，经常需要添加 User-Agent 库和 Referer 库。关于 Referer 和 User-Agent 的作用，读者可以查看第 3 章的内容。在程序 4-5 所示代码的静态类 Builder 中，userAgentList 保存了多个 User-Agent，refererList 集合保存了多个 Referer。在主方法中，对静态类 Builder 进行实例化，便可以使用类中的变量。每次请求 URL 时，User-Agent 和 Referer 的值分别从 User-Agent 库和 Referer 库中随机产生。通过添加 User-Agent 库和 Referer 库，大大减少了网络爬虫被封的风险。

程序 4-5

```java
import java.io.IOException;
import java.util.Arrays;
import java.util.HashMap;
import java.util.List;
import java.util.Map;
import java.util.Random;
import org.jsoup.Connection;
import org.jsoup.Jsoup;
import org.jsoup.nodes.Document;
public class JsoupConnectHeaderList {
  public static void main(String[] args) throws IOException {
      Connection connect = Jsoup.connect
("http://www.********.com.cn/b.asp");
      //实例化静态类
      Builder builder = new Builder();
      //请求网页添加不同Host，也可以不设置
      builder.host = "www.********.com.cn";
      //将Builder中的信息添加到Map集合中
      Map<String, String> header = new HashMap<String, String>();
      header.put("Host", builder.host);
      header.put("User-Agent",
            builder.userAgentList.get(new Random().nextInt
(builder.userAgentSize)) );
      header.put("Accept", builder.accept);
      header.put("Referer", builder.refererList.get(new Random().
nextInt(builder.refererSize)));
      header.put("Accept-Language", builder.acceptLanguage);
      header.put("Accept-Encoding", builder.acceptEncoding);
      //设置头
      Connection conheader = connect.headers(header);
      Document document = conheader.get();     //发送GET请求
      System.out.println(document);            //输出HTML
  }
  /**
   * 封装请求头信息的静态类
   */
  static class Builder{
      //设置User-Agent库；读者根据需求添加更多User-Agent
      String[] userAgentStrs = {"Mozilla/5.0 (Macintosh; U; Intel Mac OS X 10_6_8; en-us) AppleWebKit/534.50 (KHTML, like Gecko) Version/5.1 Safari/534.50",
```

```
                "Mozilla/5.0 (Windows; U; Windows NT 6.1; en-us) 
AppleWebKit/534.50 (KHTML, like Gecko) Version/5.1 Safari/534.50", 
                "Mozilla/5.0 (Windows NT 10.0; WOW64; rv:38.0) 
Gecko/20100101 Firefox/38.0", 
                "Mozilla/5.0 (Windows NT 10.0; WOW64; Trident/7.0; 
.NET4.0C; .NET4.0E; .NET CLR 2.0.50727; .NET CLR 3.0.30729; .NET CLR 
3.5.30729; InfoPath.3; rv:11.0) like Gecko", 
                "Mozilla/5.0 (compatible; MSIE 9.0; Windows NT 6.1; 
Trident/5.0)", 
                "Mozilla/4.0 (compatible; MSIE 8.0; Windows NT 6.0; 
Trident/4.0)", 
                 "Mozilla/4.0 (compatible; MSIE 7.0; Windows NT 6.0)"};
        List<String> userAgentList = Arrays.asList(userAgentStrs);
        int userAgentSize = userAgentList.size();
        //设置Referer库；读者根据需求添加更多Referer
        String[] refererStrs = {"https://www.*****.com/",
                "https://www.*****.com/",
                "http://www.****.com",
                "https://www.**.com/"};
        List<String> refererList = Arrays.asList(refererStrs);
        int refererSize = refererList.size();
        //设置accept、accept-Language及accept-Encoding
        String accept = "text/html,application/xhtml+xml, 
application/xml;q=0.9,image/webp,image/apng,*/*;q=0.8";
        String acceptLanguage = "zh-cn,zh;q=0.5";
        String acceptEncoding = "gzip, deflate";
        String host;
    }
}
```

4.1.4 提交请求参数

网页提交表单数据，涉及一系列请求参数。GET 请求的参数，是通过 URL 传递的，通常以"?key1=value1&key2=value2"的形式进行传递。POST 请求的参数，通常是放在 POST 请求的消息体中，格式一般为 JSON。例如，在某快递网站中输入快递单号，通过网络抓包获取请求信息，如图 4.6 所示，请求的方法为 GET，提交的参数有 3 个。

请求地址：http://www.*****.com/ems.php。

快递单号：EH629625211CS。

抓包获得的参数：wen:EH629625211CS action:ajax rnd: 0.15938420328106995。

图 4.6　表单提交数据

Jsoup 提供了 5 种添加请求参数的方法，如下所示。

```
Connection data(String key, String value);
Connection data(String... keyvals);
Connection data(Map<String, String> data);
Connection data(String key, String filename, InputStream inputStream);
Connection data(String key, String filename, InputStream inputStream, String contentType);
Connection data(Collection<KeyVal> data);
```

在这 5 种方法中，使用较多的是前 3 种方法。以某快递网站为例，分别演示这 3 种方法的使用，如程序 4-6、程序 4-7 和程序 4-8 所示。这 3 个程序在控制台的输出结果相同，如图 4.7 所示。另外，在这 3 个程序中实际提交的参数只有 2 个。

程序 4-6

```
Connection connect = Jsoup.connect("http://www.*****.com/ems.php");
//添加参数
connect.data("wen","EH629625211CS").data("action", "ajax");
Response response = connect.method(Method.GET).ignoreContentType(true).execute();
//获取数据，转换成HTML格式
Document document = response.parse();
System.out.println(document);
```

程序 4-7

```
Connection connect = Jsoup.connect("http://www.*****.com/ems.php");
//添加参数
Map<String, String> data = new HashMap<String, String>();
data.put("wen", "EH629625211CS");
data.put("action", "ajax");
//获取响应
Response response = connect.data(data).method(Method.GET).ignoreContentType(true).execute();
//获取数据，转换成HTML格式
Document document = response.parse();
System.out.println(document);
```

程序 4-8

```
Connection connect = Jsoup.connect("http://www.*****.com/ems.php");
//添加参数
connect.data("wen", "EH629625211CS", "action", "ajax");
Response response = connect.method(Method.GET).ignoreContentType(true).execute();
//获取数据，转换成HTML格式
Document document = response.parse();
System.out.println(document);
```

```
<html>
 <head></head>
 <body>
  <span style="color:red;">EMS快递单号【EH629625211CS】暂无信息,可尝试刷新一遍试试！！</span>
 </body>
</html>
```

图 4.7　请求该快递网站返回的响应内容

4.1.5　超时设置

在网络异常情况下，可能会发生连接超时，进而导致程序僵死而不继续往下执行。在 Jsoup 请求 URL 时，如果发生连接超时的情况，则会发出如图 4.8 所示的异常信息。

```
Exception in thread "main" java.net.ConnectException: Connection timed out: connect
    at java.net.DualStackPlainSocketImpl.waitForConnect(Native Method)
    at java.net.DualStackPlainSocketImpl.socketConnect(DualStackPlainSocketImpl.java:85)
    at java.net.AbstractPlainSocketImpl.doConnect(AbstractPlainSocketImpl.java:350)
    at java.net.AbstractPlainSocketImpl.connectToAddress(AbstractPlainSocketImpl.java:206)
    at java.net.AbstractPlainSocketImpl.connect(AbstractPlainSocketImpl.java:188)
    at java.net.PlainSocketImpl.connect(PlainSocketImpl.java:172)
    at java.net.SocksSocketImpl.connect(SocksSocketImpl.java:392)
    at java.net.Socket.connect(Socket.java:589)
    at sun.security.ssl.SSLSocketImpl.connect(SSLSocketImpl.java:668)
    at sun.net.NetworkClient.doConnect(NetworkClient.java:175)
    at sun.net.www.http.HttpClient.openServer(HttpClient.java:432)
    at sun.net.www.http.HttpClient.openServer(HttpClient.java:527)
    at sun.net.www.protocol.https.HttpsClient.<init>(HttpsClient.java:264)
    at sun.net.www.protocol.https.HttpsClient.New(HttpsClient.java:367)
    at sun.net.www.protocol.https.AbstractDelegateHttpsURLConnection.getNewHttpClient(AbstractDelegateHttpsURLConn
    at sun.net.www.protocol.http.HttpURLConnection.plainConnect0(HttpURLConnection.java:1105)
    at sun.net.www.protocol.http.HttpURLConnection.plainConnect(HttpURLConnection.java:999)
    at sun.net.www.protocol.https.AbstractDelegateHttpsURLConnection.connect(AbstractDelegateHttpsURLConnection.ja
    at sun.net.www.protocol.https.HttpsURLConnectionImpl.connect(HttpsURLConnectionImpl.java:153)
    at org.jsoup.helper.HttpConnection$Response.execute(HttpConnection.java:746)
    at org.jsoup.helper.HttpConnection$Response.execute(HttpConnection.java:722)
    at org.jsoup.helper.HttpConnection.execute(HttpConnection.java:306)
    at org.jsoup.helper.HttpConnection.get(HttpConnection.java:295)
    at com.qian.jsoupconnect.Test.main(Test.java:15)
```

图 4.8　连接超时异常

针对连接超时问题，Jsoup 在请求 URL 时，可以由用户自行设置毫秒级超时时间，如程序 4-9 所示。如果不使用 timeout 方法设置超时时间，则超时时间默认为 30 毫秒。

程序 4-9

```
//基于 timeout 设置超时时间
Response response = Jsoup.connect("https://******.com/")
        .method(Method.GET).timeout(3*1000).execute();
Document document = Jsoup.connect("https://******.com/").
timeout(10*1000).get();
```

4.1.6　代理服务器的使用

在一般情况下，使用浏览器可以直接连接 Internet 站点获取网络信息，而代理服务器（Proxy Server）是网络上提供转接功能的服务器。代理服务器是介于客户端和 Web 服务器之间的另一台服务器，基于代理服务器，浏览器不再直接从 Web 服务器获取数据，而是向代理服务器发出请求，信号会先送发到代理服务器，由代理服务器取回浏览器所需要的信息。可以将代理简单理解为中介。

在网络爬虫中，使用代理服务器访问网页内容，能够高度隐藏爬虫的真实 IP 地址，从而防止网络爬虫被服务器封锁。另外，普通网络爬虫使用固定 IP 地址请求时，往往需要设置随机休息时间，而通过代理服务器却不需要，从而提高了数据采集的效率。目前，代理服务器可以来源于提供免费代理服务的一些网站或接口网站，但这些免费代理 IP 地址的稳定性较差。另外，也可通过付费的方式获取商业级代理，其提供的代理 IP 地址可用率较高，稳定性较强。

在 Jsoup 中，提供了两种方式设置代理服务器，如下所示。

```
/**
 * Set the proxy to use for this request. Set to <code>null</code>
```

```
to disable.
     * @param proxy proxy to use  代理
     * @return this Connection, for chaining  返回值为Connection
     */
    Connection proxy(Proxy proxy);
    /**
     * Set the HTTP proxy to use for this request.
     * @param host the proxy hostname  IP地址
     * @param port the proxy port       端口
     * @return this Connection, for chaining  返回值为Connection
     */
    Connection proxy(String host, int port);
```

仍以请求某培训网站的网址为例,演示这两种设置代理服务器的方法,如程序4-10 和程序 4-11 所示。在设置代理服务器时,需要知道代理服务器的 IP 地址以及端口。在本案例中,只使用了一个代理服务器请求网页,代理服务器的 IP 地址为"171.221.239.11",端口为 808。在实际应用中,往往需要构建代理服务器库,不断地切换代理服务器去请求 URL 库。

程序 4-10

```java
import java.io.IOException;
import java.net.InetSocketAddress;
import java.net.Proxy;
import org.jsoup.*;
import org.jsoup.Connection.*;
public class JsoupConnectProxy1 {
  public static void main(String[] args) throws IOException {
      //使用第一种方式设置代理
      Proxy proxy = new Proxy(Proxy.Type.HTTP, new InetSocketAddress("171.221.239.11", 808));
      Connection connection = Jsoup.connect("http://www.********.com.cn/b.asp").proxy(proxy);
      Response response = connection.method(Method.GET).timeout(10*1000).execute();
      //获取响应状态码
      int statusCode = response.statusCode();
      System.out.println("响应状态码为:" + statusCode);
  }
}
```

程序 4-11

```
//使用第二种方式设置代理
Connection connection = Jsoup.connect
("http://www.********.com.cn/b.asp")
     .proxy("171.221.239.11",808);
Response response = connection.method(Method.GET).timeout (10*1000).
execute();
//获取响应状态码
int statusCode = response.statusCode();
System.out.println("响应状态码为:" + statusCode);
```

4.1.7 响应转输出流（图片、PDF 等的下载）

使用 Jsoup 下载图片、PDF 和压缩等文件时，需要将响应转化成输出流。转化成输出流的目的是增强写文件的能力，即以字节为单位写入指定文件。关于输出流的详细内容，将在第 6 章进行介绍。

以图片下载为例，程序 4-12 使用 bodyStream() 方法将响应转化成输出流，并以缓冲流的方式写入指定文件。另外，针对图片和 PDF 等文件，在执行 URL 请求获取 Response 时，必须通过 ignoreContentType(boolean ignoreContentType) 方法设置忽略响应内容的类型，否则会报错。

程序 4-12

```java
import java.io.BufferedInputStream;
import java.io.BufferedOutputStream;
import java.io.File;
import java.io.FileOutputStream;
import java.io.IOException;
import org.jsoup.Connection;
import org.jsoup.Jsoup;
import org.jsoup.Connection.Method;
import org.jsoup.Connection.Response;
public class JsoupConnectInputtstream {
  public static void main(String[] args) throws IOException {
      String imageUrl = "http://i-4.******.com/2018/6/11/KDE5Mngp/
                ae0c2d4d-04fb-4066-872c-a8c7a7c4ea4f.jpg";
      Connection connect = Jsoup.connect(imageUrl);
      Response response = connect.method(Method.GET).
ignoreContentType(true).execute();
```

```java
        System.out.println("文件类型为:" + response.contentType());
        //响应转化成输出流
        BufferedInputStream bufferedInputStream = response.bodyStream();
        //保存图片
        saveImage(bufferedInputStream,"image/1.jpg");
    }
    /**
     * 保存图片操作
     * @param 输入流
     * @param 保存的文件目录
     * @throws IOException
     */
    static void saveImage(BufferedInputStream in, String savePath) throws IOException {
        byte[] buffer = new byte[1024];
        int len = 0;
        //创建缓冲流
        FileOutputStream fileOutStream = new FileOutputStream(new File(savePath));
        BufferedOutputStream bufferedOut = new BufferedOutputStream(fileOutStream);
        //图片写入
        while ((len = in.read(buffer, 0, 1024)) != -1) {
            bufferedOut.write(buffer, 0, len);
        }
        //缓冲流释放与关闭
        bufferedOut.flush();
        bufferedOut.close();
    }
}
```

运行程序 4-12，将在控制台输出响应文件的类型，如图 4.9 所示。由图 4.10 可以看出，图片被成功下载到了程序指定的目录。

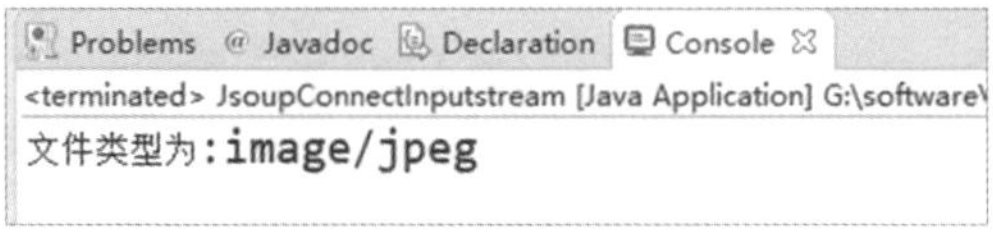

图 4.9　程序 4-12 执行的结果

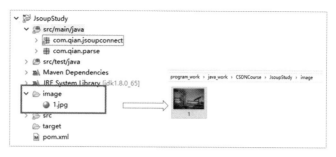

图 4.10　图片下载结果

4.1.8　HTTPS 请求认证

以 https://为前缀的 URL 使用的是 HTTPS 协议。HTTPS 在 HTTP 的基础上加入了 SSL（安全套接层）。SSL 的作用是保障网络通信的安全性，其广泛应用于客户端与服务器之间的身份认证和加密数据传输。

SSL 支持双向认证（服务器认证与客户端认证），将服务器证书下载到客户端，再将客户端的证书返回到服务器。目前，访问网站并不常用客户端证书，大部分用户都没有自己的客户端证书，但 HTTPS 总要求使用客户端证书。其中，使用最多的客户端证书是 X.509 证书。

网络爬虫在请求以 https://为前缀的 URL 时，通常也需要创建 X.509 证书信任管理器。在程序 4-13 中，使用 Jsoup 请求以 https://为前缀的 URL，没有创建证书信任管理器。运行程序 4-13，会出现如图 4.11 所示的错误信息，即找不到合法的证书去请求目标 URL。

程序 4-13

```
Connection connect = Jsoup.connect("https://cn.*******.com");
Document document = connect.get();
System.out.println(document);
```

```
Exception in thread "main" javax.net.ssl.SSLHandshakeException:
sun.security.validator.ValidatorException: PKIX path building failed:
sun.security.provider.certpath.SunCertPathBuilderException: unable to find
valid certification path to requested target ...
```

图 4.11　运行程序 4-13 的报错内容

查看 Jsoup 源码，发现在 org.jsoup.helper.HttpConnection 类中提供实现信任管理器

的 initUnSecureTSL() 方法，如程序 4-14 所示。使用这种方法，可以成功请求以 https:// 为前缀的 URL。

程序 4-14

```java
/**
 * Initialise Trust manager that does not validate certificate chains and
 * add it to current SSLContext.
 * <p/>
 * please not that this method will only perform action if sslSocketFactory is not yet
 * instantiated.
 * 初始化信任管理器，不验证证书
 * @throws IOException on SSL init errors
 */
private static synchronized void initUnSecureTSL() throws IOException {
    if (sslSocketFactory == null) {
        final TrustManager[] trustAllCerts = new TrustManager[]{new X509TrustManager() {

            public void checkClientTrusted(final X509Certificate[] chain, final String authType) {
            }

            public void checkServerTrusted(final X509Certificate[] chain, final String authType) {
            }

            public X509Certificate[] getAcceptedIssuers() {
                return null;
            }
        }};
        final SSLContext sslContext;
        try {
            sslContext = SSLContext.getInstance("SSL");
            sslContext.init(null, trustAllCerts, new java.security.SecureRandom());
            sslSocketFactory = sslContext.getSocketFactory();
        } catch (NoSuchAlgorithmException | KeyManagementException e) {
```

```
                throw new IOException("Can't create unsecure trust
manager");
            }
        }
    }
```

但仔细阅读 org.jsoup.helper.HttpConnection 类中的 createConnection()方法对应的源码时，发现 initUnSecureTSL()方法的调用是需要一定条件的，如程序 4-15 所示。由程序 4-15 可知，只有在 req.validateTLSCertificates()执行的结果为 false 的情况下，才会执行 initUnSecureTSL()方法。而 validateTLSCertificates()方法实际返回的是静态类 Request 中的 validateTSLCertificates 成员变量(默认情况下为 true)，如程序 4-16 所示。

程序 4-15

```
    if (conn instanceof HttpsURLConnection) {
        SSLSocketFactory socketFactory = req.sslSocketFactory();
        if (socketFactory != null) {
            ((HttpsURLConnection) conn).setSSLSocketFactory
(socketFactory);
        } else if (!req.validateTLSCertificates()) {
            initUnSecureTSL();

   ((HttpsURLConnection)conn).setSSLSocketFactory(sslSocketFactory);
        ((HttpsURLConnection)conn).setHostnameVerifier
(getInsecureVerifier());
        }
    }
```

程序 4-16

```
    private boolean validateTSLCertificates = true;

    public boolean validateTSLCertificates() {
        return validateTSLCertificates;
    }
```

因此，需要将 validateTSLCertificates 的值设置为 false 才能调用 initUnSecureTSL()方法。在 org.jsoup.helper.HttpConnection 类中提供了设置 validateTSLCertificates 值的方法，如程序 4-17 所示。

程序 4-17

```
public Connection validateTLSCertificates(boolean value) {
    req.validateTLSCertificates(value);
    return this;
}
```

为此，可以将程序 4-13 修改为程序 4-18，便可以成功请求以 https:// 为前缀的 URL。

程序 4-18

```
Connection connect = Jsoup.connect("https://cn.*******.com")
    .validateTLSCertificates(false);
Document document = connect.get();
System.out.println(document);
```

另外，也可以在 Jsoup 创建 Connection 连接之前，调用自己编写的创建信任管理器（不验证证书）的方法，如程序 4-19 所示。基于程序 4-19 也可以成功地请求以 https:// 为前缀的 URL。

程序 4-19

```java
import java.io.IOException;
import java.security.cert.X509Certificate;
import javax.net.ssl.HttpsURLConnection;
import javax.net.ssl.SSLContext;
import javax.net.ssl.SSLSocketFactory;
import javax.net.ssl.TrustManager;
import javax.net.ssl.X509TrustManager;
import org.jsoup.Connection;
import org.jsoup.Jsoup;
import org.jsoup.nodes.Document;
public class JsoupConnectSSLInit {
  public static void main(String[] args) throws IOException {
      initUnSecureTSL();
      String url = "https://cn.*******.com";
      //创建连接
      Connection connect = Jsoup.connect(url);
      //请求网页
      Document document = connect.get();
      //输出HTML
      System.out.println(document.html());
  }
```

```java
    private static void initUnSecureTSL() {
        // 创建信任管理器(不验证证书)
        final TrustManager[] trustAllCerts = new TrustManager[]{new X509TrustManager() {
            //检查客户端证书
            public void checkClientTrusted(final X509Certificate[] chain, final String authType) {
                //do nothing 接受任意客户端证书
            }
            //检查服务器端证书
            public void checkServerTrusted(final X509Certificate[] chain, final String authType) {
                //do nothing  接受任意服务端证书
            }
            //返回受信任的X509证书
            public X509Certificate[] getAcceptedIssuers() {
                return null;
            }
        }};
        try {
            // 创建SSLContext对象,并使用指定的信任管理器初始化
            SSLContext sslContext = SSLContext.getInstance("SSL");
            sslContext.init(null, trustAllCerts, new java.security.SecureRandom());
            //基于信任管理器,创建套接字工厂
            SSLSocketFactory sslSocketFactory = sslContext.getSocketFactory();
            //为HttpsURLConnection配置套接字工厂
            HttpsURLConnection.setDefaultSSLSocketFactory(sslSocketFactory);
        } catch (Exception e) {
            e.printStackTrace();
        }
    }
}
```

4.1.9 大文件内容获取问题

在采集数据时,经常会遇到一些较大的文件,如包含大量文本信息的 HTML 文件、大小超过 10M 的图片、PDF 和 ZIP 等文件。在默认情况下,Jsoup 最大只能获取 1M

的文件。因此，直接使用 Jsoup 请求包含大量文本信息的 HTML 文件，将导致获取的内容不全；请求下载大小超过 1M 的图片和 ZIP 等文件,将导致文件无法查看或解压。但在 Jsoup 中，可以使用 maxBodySize(int bytes)设置请求文件大小限制，来避免这种问题的出现。例如，使用程序 4-20 下载 httpd-2.4.37.tar.gz 文件（大小为 8.75M）时，Integer.MAX_VALUE 为设置的请求文件大小。另外，在请求大文件时，设置的超时时间也需尽量长些。

程序 4-20

```java
import java.io.BufferedInputStream;
import java.io.BufferedOutputStream;
import java.io.File;
import java.io.FileOutputStream;
import java.io.IOException;
import org.jsoup.Jsoup;
import org.jsoup.Connection.Method;
import org.jsoup.Connection.Response;
public class JsoupConnectBodySize1 {
  public static void main(String[] args) throws IOException {
        String url = "https://www-us.******.org/dist//httpd/httpd-2.4.37.tar.gz";
        //超时时间设置长一些，下载大文件
        Response response = Jsoup.connect(url).timeout(10*60*1000)
                .maxBodySize(Integer.MAX_VALUE)
                .method(Method.GET).ignoreContentType(true).execute();
        //如果响应成功，则执行下面的操作
        if (response.statusCode() == 200) {
            //响应转化成输出流
            BufferedInputStream bufferedInputStream = response.bodyStream();
            //保存图片
            saveFile(bufferedInputStream,"image/httpd-2.4.37.tar.gz");
        }
    }
    /**
     * 保存文件
     * @param 输入流
     * @param 保存的文件目录
     * @throws IOException
     */
```

```java
    static void saveFile(BufferedInputStream inputStream, String savePath) throws IOException {
        //一次最多读取1KB
        byte[] buffer = new byte[1024];
        int len = 0;
        //创建缓冲流
        FileOutputStream fileOutStream = new FileOutputStream(new File(savePath));
        BufferedOutputStream bufferedOut = new BufferedOutputStream(fileOutStream);
        //文件写入
        while ((len = inputStream.read(buffer, 0, 1024)) != -1) {
            bufferedOut.write(buffer, 0, len);
        }
        //缓冲流释放与关闭
        bufferedOut.flush();
        bufferedOut.close();
    }
}
```

4.2 HttpClient 的使用

HttpClient 是 Apache Jakarta Common 下的子项目，用来提供高效的、功能丰富的、支持 HTTP 协议的客户端编程工具包。相比于 java.net 包中提供的 URLConnection 与 HttpURLConnection，HttpClient 增加了易用性和灵活性。在 Java 网络爬虫实战中，经常使用 HttpClient 向服务器发送请求，获取响应资源。官网提供了 HttpClient 的使用教程。

HttpClient 涉及的内容较多，本节主要介绍 HttpClient 的基本用法。关于其他内容，读者可以基于 HttpClient 源码深入学习。

4.2.1 jar 包的下载

首先，在 Eclipse 中创建 Maven 工程，并在 Maven 工程的 pom.xml 文件中添加 HttpClient 对应的 dependency。

```
<!-- https://*************.com/artifact/org.apache.httpcomponents/httpclient -->
    <dependency>
        <groupId>org.apache.httpcomponents</groupId>
```

```
            <artifactId>httpclient</artifactId>
            <version>4.5.5</version>
</dependency>
```

基于 pom.xml 文件的配置信息,可以下载 HttpClient 的 jar 包(版本是 4.5.5)。另外,可以看到 HttpClient 的依赖 jar 包有 commons-codec、commons-logging 和 httpcore,如图 4.12 所示。

图 4.12　HttpClient 的 jar 包及其相关依赖 jar 包

4.2.2　请求 URL

1. 创建 HttpClient 实例

HttpClient 的重要功能是执行 HTTP 请求方法,获取响应资源。在执行具体的请求方法之前,需要实例化 HttpClient。程序 4-21 提供了实例化 HttpClient 的六种方式,第一种方式已经不再建议使用。

程序 4-21

```
        /*HttpClient 实例化方法
        *HttpClients.custom()返回值HttpClientBuilder.create()
        *HttpClients.createDefault()返回值 HttpClients.custom().build()
```

```
*具体可阅读HttpClients类
**/
HttpClient httpClient1 = new DefaultHttpClient();
HttpClient httpClient2 = HttpClients.custom().build();
HttpClient httpClient3 = HttpClientBuilder.create().build();
CloseableHttpClient httpClient4 = HttpClients.createDefault();
HttpClient httpClient5 = HttpClients.createSystem();
HttpClient httpClient6 = HttpClients.createMinimal();
```

2. 创建请求方法实例

在HttpClient中，支持HTTP/1.1版本中的所有HTTP方法，即GET、POST、HEAD、PUT、DELETE、OPTIONS 和 TRACE。其中，每种方法都对应一个类，即HttpGet、HttpPost、HttpHead、HttpPut、HttpDelete、HttpOption 和 HttpTrace。在网络爬虫中，常用的类是HttpGet 与 HttpPost。从HttpClient 源码中，可以发现这些类的实例化方式各有三种（以HttpGet 为例），如程序4-22所示。

程序 4-22

```java
public HttpGet() {
    super();
}
public HttpGet(final URI uri) {
    super();
    setURI(uri);
}
public HttpGet(final String uri) {
    super();
    setURI(URI.create(uri));
}
```

其中，采用第一种方式实例化，还需要设置请求的URL；第二种方式输入参数是统一资源标识符URI，第三种方式输入参数是字符串类型的URI。程序4-23演示了这三种方式的使用情况，程序的输出结果如图4.13所示。

程序 4-23

```java
import java.net.URI;
import java.net.URISyntaxException;
import org.apache.http.client.methods.HttpGet;
import org.apache.http.client.utils.URIBuilder;
public class HttpGetInit {
```

```java
public static void main(String[] args) throws URISyntaxException {
    //第一种方式
    String personalUrl = "http://www.********.com.cn/b.asp";
    URI uri = new URIBuilder(personalUrl).build();    //创建URI
    HttpGet getMethod = new HttpGet(); //GET方法请求
    getMethod.setURI(uri);                   //设置URI
    System.out.println(getMethod);
    //第二种方式
    HttpGet httpGetUri = new HttpGet(uri);
    System.out.println(httpGetUri);
    //第三种方式
    HttpGet httpGetStr = new HttpGet(personalUrl);
    System.out.println(httpGetStr);
}
```

```
GET http://www.▇▇▇▇▇.com.cn/b.asp HTTP/1.1
GET http://www.▇▇▇▇▇.com.cn/b.asp HTTP/1.1
GET http://www.▇▇▇▇▇.com.cn/b.asp HTTP/1.1
```

图 4.13 程序 4-23 的输出结果

3. 执行请求

基于实例化的 HttpClient，可以调用 execute(HttpUriRequest request)方法执行数据请求，返回 HttpResponse。程序 4-24、程序 4-25 和程序 4-26 各给出了一种操作方式。

程序 4-24

```java
//执行请求
HttpResponse response = new BasicHttpResponse(HttpVersion.HTTP_1_1,
        HttpStatus.SC_OK, "OK");         //实例化HttpResponse
response = httpClient.execute(getMethod);       //执行响应
```

程序 4-25

```java
//执行请求
HttpResponse httpResponse = null;
try {
    httpResponse = httpClient.execute(httpGet);      //执行响应
} catch (IOException e) {
    e.printStackTrace();
}
```

程序 4-26

```java
//执行请求 CloseableHttpResponse 继承于 HttpResponse
    CloseableHttpResponse response = null;
try {
    response = httpClient.execute(httpGet);
} catch (IOException e) {
    e.printStackTrace();
}
```

另外，在 HttpClient 类中，还提供了其他执行请求的方法，以下列举了三种。

```java
//方法1：HttpContext 为 HTTP 执行环境
HttpResponse execute(HttpUriRequest request, HttpContext context)
        throws IOException, ClientProtocolException;
//方法2：HttpHost代表代理连接
HttpResponse execute(HttpHost target, HttpRequest request)
        throws IOException, ClientProtocolException;
//方法3：HttpHost代表代理连接
HttpResponse execute(HttpHost target, HttpRequest request,
HttpContext context)
        throws IOException, ClientProtocolException;
```

4．获取响应信息

基于上述方法 3 获取的 HttpResponse，可以继续执行一些方法获取响应状态码、协议版本、响应头和响应实体等信息，如程序 4-27 所示。程序 4-27 在执行请求时，使用了 HttpContext，即 HTTP 上下文环境。

程序 4-27

```java
import java.io.IOException;
import org.apache.http.Header;
import org.apache.http.HttpEntity;
import org.apache.http.HttpResponse;
import org.apache.http.HttpStatus;
import org.apache.http.ParseException;
import org.apache.http.ProtocolVersion;
import org.apache.http.client.HttpClient;
import org.apache.http.client.methods.HttpGet;
import org.apache.http.impl.client.HttpClients;
import org.apache.http.protocol.BasicHttpContext;
```

```java
import org.apache.http.protocol.HttpContext;
import org.apache.http.util.EntityUtils;
public class HttpclientTest {
  public static void main(String[] args) throws ParseException,
IOException {
        //初始化HttpContext
        HttpContext localContext = new BasicHttpContext();
        String url = "http://www.********.com.cn/b.asp";
        //初始化HttpClient
        HttpClient httpClient = HttpClients.custom().build();
        HttpGet httpGet = new HttpGet(url);
        //执行请求获取HttpResponse
        HttpResponse httpResponse = null;
        try {
            httpResponse = httpClient.execute(httpGet,localContext);
        } catch (IOException e) {
            e.printStackTrace();
        }
        //获取具体响应信息
        System.out.println("response:" + httpResponse );
        //响应状态
        String status = httpResponse .getStatusLine().toString();
        System.out.println("status:" + status);
        //获取响应状态码
        int StatusCode = httpResponse .getStatusLine().getStatusCode();
        System.out.println("StatusCode:" + StatusCode);
        ProtocolVersion protocolVersion = httpResponse .
getProtocolVersion(); //协议的版本号
        System.out.println("protocolVersion:" + protocolVersion);
        //是否OK
        String phrase = httpResponse .getStatusLine().getReasonPhrase();
        System.out.println("phrase:" + phrase);
        Header[] headers = httpResponse.getAllHeaders();
        System.out.println("输出头信息为：");
        for (int i = 0; i < headers.length; i++) {
            System.out.println(headers[i]);
        }
        System.out.println("头信息输出结束");
        if(StatusCode == HttpStatus.SC_OK){     //状态码200表示响应成功
            //获取实体内容
            HttpEntity entity = httpResponse.getEntity();
            //注意设置编码
```

```
            String entityString = EntityUtils.toString (entity,"gbk");
            //输出实体内容
            System.out.println(entityString);
            EntityUtils.consume(httpResponse.getEntity());    //消耗实体
        }else {
            //关闭HttpEntity的流实体
            EntityUtils.consume(httpResponse.getEntity());    //消耗实体
        }
    }
}
```

图 4.14 为程序 4-27 的部分输出结果。

```
response:HttpResponseProxy{HTTP/1.1 200 OK [Cache-Control: private, Content-Type: text/html, Vary: Accept-Encoding,
status:HTTP/1.1 200 OK
StatusCode:200
protocolVersion:HTTP/1.1
phrase:OK
输出头信息为:
Cache-Control: private
Content-Type: text/html
Vary: Accept-Encoding
Server: Microsoft-IIS/10.0
Set-Cookie: ASPSESSIONIDCCDCBATC=CIMGCOJCKKAAFFKDALBOFIFC; path=/
X-Powered-By: ASP.NET
Date: Tue, 23 Oct 2018 07:30:04 GMT
头信息输出结束
<!DOCTYPE html PUBLIC "-//W3C//DTD XHTML 1.0 Strict//EN" "http://www.■.org/TR/xhtml1/DTD/xhtml1-strict.dtd">
<html xmlns="http://www.■.org/1999/xhtml">
<head>
<meta http-equiv="Content-Type" content="text/html; charset=gb2312" />

<title>浏览器脚本教程</title>

<link rel="stylesheet" type="text/css" href="/c5.css" />
<link rel="shortcut icon" href="/favicon.ico" type="image/x-icon" />

</head>
```

图 4.14　程序 4-27 的部分输出结果

4.2.3　EntityUtils 类

在程序 4-27 中，使用了 HttpClient 中的 EntityUtils 类。EntityUtils 类的作用是操作响应实体。例如，数据类型为 HTML 响应实体，可以使用以下三种方法将其直接转化成字符串类型。

```
//可以设置编码
public static String toString(final HttpEntity entity, final String defaultCharset)
//可以设置编码
public static String toString(final HttpEntity entity, final Charset defaultCharset)
//使用默认编码ISO-8859-1
public static String toString(final HttpEntity entity)
```

另外，EntityUtils 类还提供了将响应实体转化成字节数组的方法，如下。

public static byte[] toByteArray(**final** HttpEntity entity)

针对图片、PDF 和压缩包等文件，可以先将响应实体转化成字节数组。之后，利用缓冲流的方式写入指定文件，具体操作程序将在第六章进行介绍。

另外，为确保系统资源的释放，可以调用下面的方法消耗实体。

public static void consume(**final** HttpEntity entity)

4.2.4 设置头信息

在 4.1.3 节中，已经介绍了网络爬虫设置头信息的作用。在本小节中，将主要介绍 HttpClient 工具如何设置请求头。仍以某培训网站网址为例，基于网络抓包获取请求头信息，如图 4.2 所示。程序 4-28 所示为 HttpClient 设置请求头的一种方式。

程序 4-28

```
        //初始化httpClient
        HttpClient httpClient = HttpClients.custom().build();
        //使用的请求方法
        HttpGet httpget = new HttpGet("http://www.********.com.cn/b.asp");
        //请求头配置
        httpget.setHeader("Accept", "text/html,application/xhtml+xml,application/xml;q=0.9,image/webp,image/apng,*/*;q=0.8");
        httpget.setHeader("Accept-Encoding", "gzip, deflate");
        httpget.setHeader("Accept-Language", "zh-CN,zh;q=0.9");
        httpget.setHeader("Cache-Control", "max-age=0");
        httpget.setHeader("Host", "www.********.com.cn");
        httpget.setHeader("User-Agent", "Mozilla/5.0 (Windows NT 10.0; Win64; x64) AppleWebKit/537.36 (KHTML, like Gecko) Chrome/63.0.3239.108 Safari/537.36"); //这项内容很重要
        //发出GET请求
        HttpResponse response = httpClient.execute(httpget);
        //获取响应状态码
        int code = response.getStatusLine().getStatusCode();
        HttpEntity httpEntity = response.getEntity();  //获取网页内容流
        //以字符串的形式(需设置编码)
        String entity = EntityUtils.toString(httpEntity, "gbk");
```

```java
        System.out.println(code + "\n" + entity);    //输出所获得的内容
        EntityUtils.consume(httpEntity);             //关闭内容流
```

程序4-29所示为设置请求头的另一种方式。

程序4-29

```java
    //通过集合封装头信息
    List<Header> headerList = new ArrayList<Header>();
    headerList.add(new BasicHeader(HttpHeaders.ACCEPT, "text/html,application/xhtml+xml,application/xml;q=0.9,image/webp,image/apng,*/*;q=0.8"));
    headerList.add(new BasicHeader(HttpHeaders.USER_AGENT, "Mozilla/5.0 (Windows NT 10.0; Win64; x64) AppleWebKit/537.36 (KHTML, like Gecko) Chrome/63.0.3239.108 Safari/537.36"));
    headerList.add(new BasicHeader(HttpHeaders.ACCEPT_ENCODING, "gzip, deflate"));
    headerList.add(new BasicHeader(HttpHeaders.CACHE_CONTROL, "max-age=0"));
    headerList.add(new BasicHeader(HttpHeaders.CONNECTION, "keep-alive"));
    headerList.add(new BasicHeader(HttpHeaders.ACCEPT_LANGUAGE, "zh-CN,zh;q=0.9"));
    headerList.add(new BasicHeader(HttpHeaders.HOST, "www.********.com.cn"));
        //构造自定义的HttpClient对象
        HttpClient httpClient = HttpClients.custom()
                    .setDefaultHeaders(headerList).build();
    //使用的请求方法
    HttpGet httpget = new HttpGet("http://www.********.com.cn/b.asp");
    //获取结果
    //发出GET请求
    HttpResponse response = httpClient.execute(httpget);
    //获取响应状态码
    int code = response.getStatusLine().getStatusCode();
    HttpEntity httpEntity = response.getEntity();    //获取网页内容流
    //以字符串的形式(需设置编码)
    String entity = EntityUtils.toString(httpEntity, "gbk");
    System.out.println(code + "\n" + entity);        //输出所获得的内容
    EntityUtils.consume(httpEntity);                 //关闭内容流
```

另外，也可以参照程序 4-5 中的静态类 Builder 构建 User-Agent 库和 Referer 库。每次请求 URL 时，User-Agent 和 Referer 的值分别从 User-Agent 库和 Referer 库中随机产生。

4.2.5　POST 提交表单

在网络爬虫中，经常遇到表单的提交，尤其是模拟登录。在 HttpClient 中，提供了实体类 UrlEncodedFormEntity 以方便处理表单提交，其使用方式如程序 4-30 所示。

程序 4-30

```java
//建立NameValuePair数组，用于存储欲传送的参数
List<NameValuePair> nvps= new ArrayList<NameValuePair>();
nvps.add(new BasicNameValuePair("param1", "value1"));
nvps.add(new BasicNameValuePair("param2", "value2"));
UrlEncodedFormEntity entity = new UrlEncodedFormEntity(nvps, Consts.UTF_8);
HttpPost httppost = new HttpPost("http://localhost/handler.do");
httppost.setEntity(entity);
```

UrlEncodedFormEntity 实例将会使用 URL encoding 来编码参数，产生如下所示的内容。

```
param1=value1&param2=value2
```

下面，以模拟登录某网站为例讲解如何使用该方法。由图 4.15 可以看出，只有提交正确的用户名及密码才能完成登录，进而访问登录后的其他页面。

图 4.15　某网站登录页面

为分析登录时，客户端向服务器提交什么参数，这里需要利用谷歌浏览器进行网络抓包，网络抓包结果如图 4.16 所示。从图 4.16 中可以看到具体的请求地址、请求方法、响应状态码和 POST 提交的参数等信息。

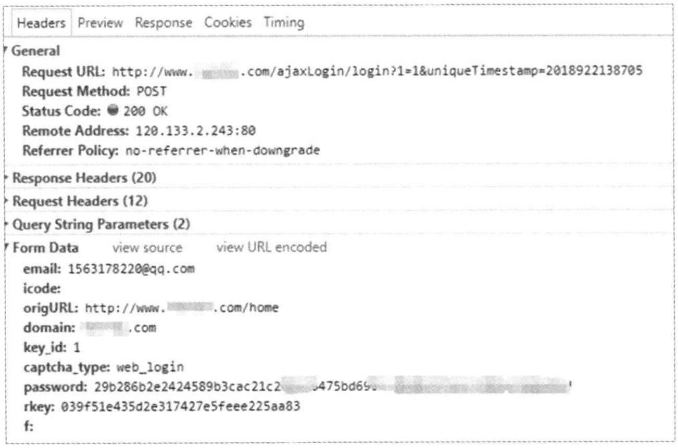

图 4.16 某网站登录网络抓包结果

基于网络抓包结果，可以模拟表单向服务器发送请求（需要提交的核心参数是用户名和密码），实现模拟登录。在登录完成后，便可以请求该网站的其他页面，如请求好友个人信息页面。完整的操作内容如程序 4-31 所示。

程序 4-31

```java
import java.io.IOException;
import java.util.ArrayList;
import java.util.List;
import org.apache.http.HttpResponse;
import org.apache.http.NameValuePair;
import org.apache.http.ParseException;
import org.apache.http.client.HttpClient;
import org.apache.http.client.ResponseHandler;
import org.apache.http.client.entity.UrlEncodedFormEntity;
import org.apache.http.client.methods.HttpGet;
import org.apache.http.client.methods.HttpPost;
import org.apache.http.impl.client.BasicResponseHandler;
import org.apache.http.impl.client.HttpClients;
import org.apache.http.message.BasicNameValuePair;
import org.apache.http.protocol.HTTP;
import org.apache.http.util.EntityUtils;
public class HttpclientRenren {
    public static void main(String[] args) throws ParseException,
```

```java
IOException {
    //实例化HttpClient
    HttpClient httpclient = HttpClients.custom().build();
    String renRenLoginURL = "http://www.******.com/ajaxLogin/login?1=1&uniqueTimestamp=2018922138705";   //登录的地址
    //采用的方法为POST
    HttpPost httppost = new HttpPost(renRenLoginURL);
    //建立NameValuePair数组,用于存储欲传送的参数
    List<NameValuePair> nvps = new ArrayList<NameValuePair>();
    //输入你的邮箱地址
    nvps.add(new BasicNameValuePair("email", ""));
    //输入你的密码
    nvps.add(new BasicNameValuePair("password", ""));
    HttpResponse response = null;
    try {
        //表单参数提交
        httppost.setEntity(new UrlEncodedFormEntity(nvps, HTTP.UTF_8));
        response = httpclient.execute(httppost);
    } catch (Exception e) {
        e.printStackTrace();
    } finally {
        httppost.abort();      //释放连接
    }
    System.out.println(response.getStatusLine());
    String entityString = EntityUtils.toString (response.getEntity(),"gbk");          //注意设置编码
    System.out.println(entityString);
    //登录完成之后需要请求的页面,这里为个人好友的信息页面
    HttpGet httpget = new HttpGet("http://www.******.com/465530468/profile?v=info_timeline");
    //构建 ResponseHandler
    ResponseHandler<String> responseHandler = new BasicResponseHandler();
    String responseBody = "";
    try {
        responseBody = httpclient.execute(httpget, responseHandler);
    } catch (Exception e) {
        e.printStackTrace();
        responseBody = null;
    } finally {
        httpget.abort();                          //释放连接
```

```
        }
        System.out.println(responseBody); //输出请求到的内容
    }
}
```

图 4.17 所示为程序 4-31 的执行结果，由该图可以发现程序成功提交了表单参数，并且模拟登录成功。

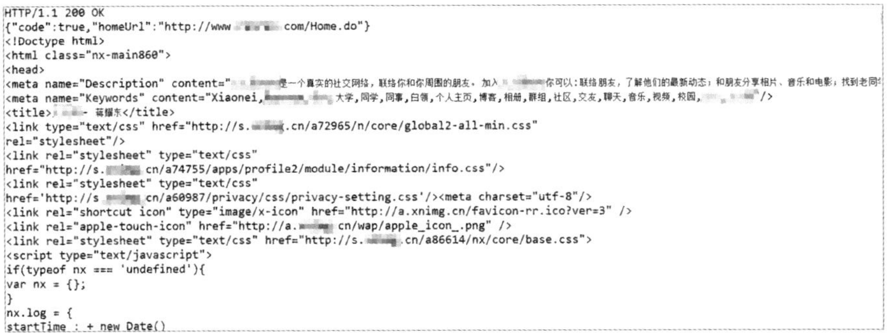

```
HTTP/1.1 200 OK
{"code":true,"homeUrl":"http://www com/Home.do"}
<!Doctype html>
<html class="nx-main860">
<head>
<meta name="Description" content="是一个真实的社交网络，联络你和你周围的朋友。加入你可以：联络朋友，了解他们的最新动态；和朋友分享相片、音乐和电影；找到老同
<meta name="Keywords" content="Xiaonei, 大学,同学,同事,白领,个人主页,博客,相册,群组,社区,交友,聊天,音乐,视频,校园,"/>
<title>- 蒋耀东</title>
<link type="text/css" href="http://s..cn/a72965/n/core/global2-all-min.css"
rel="stylesheet"/>
<link rel="stylesheet" type="text/css"
href="http://s. cn/a74755/apps/profile2/module/information/info.css"/>
<link rel="stylesheet" type="text/css"
href='http://s cn/a60987/privacy/css/privacy-setting.css'/><meta charset="utf-8"/>
<link rel="shortcut icon" type="image/x-icon" href="http://a.xnimg.cn/favicon-rr.ico?ver=3" />
<link rel="apple-touch-icon" href="http://a. cn/wap/apple_icon_.png" />
<link rel="stylesheet" type="text/css" href="http://s..cn/a86614/nx/core/base.css">
<script type="text/javascript">
if(typeof nx === 'undefined'){
var nx = {};
}
nx.log = {
startTime : + new Date()
```

图 4.17　程序 4-31 的执行结果

4.2.6　超时设置

使用 HttpClient 可配置三种超时时间：RequestTimeout（获取连接超时时间）、ConnectTimeout（建立连接超时时间）、SocketTimeout（获取数据超时时间）。配置这三种超时时间，需要用到 HttpClient 的 RequestConfig 类中的方法 custom()，该方法返回值为实例化的内部类 Builder（配置器），如程序 4-32 所示。内部类 Builder 的功能是配置相关请求的字段，如程序 4-33 所示。由程序 4-33 可以发现 Builder 不仅可以配置超时时间，还可以配置代理（proxy）、Cookie 规范（cookieSpec）、是否允许 HTTP 相关认证等。

程序 4-32

```
public static RequestConfig.Builder custom() {
    return new Builder();
}
```

程序 4-33

```
private boolean expectContinueEnabled;
private HttpHost proxy;
private InetAddress localAddress;
private boolean staleConnectionCheckEnabled;
```

```java
        private String cookieSpec;
        private boolean redirectsEnabled;
        private boolean relativeRedirectsAllowed;
        private boolean circularRedirectsAllowed;
        private int maxRedirects;
        private boolean authenticationEnabled;
        private Collection<String> targetPreferredAuthSchemes;
        private Collection<String> proxyPreferredAuthSchemes;
        private int connectionRequestTimeout;
        private int connectTimeout;
        private int socketTimeout;
        private boolean contentCompressionEnabled;

        Builder() {
            super();
            this.staleConnectionCheckEnabled = false;
            this.redirectsEnabled = true;
            this.maxRedirects = 50;
            this.relativeRedirectsAllowed = true;
            this.authenticationEnabled = true;
            this.connectionRequestTimeout = -1;
            this.connectTimeout = -1;
            this.socketTimeout = -1;
            this.contentCompressionEnabled = true;
        }
```

以下将使用 RequestConfig 类配置超时时间，操作方式如程序 4-34 所示。

程序 4-34

```java
        //全部设置为10秒
        RequestConfig requestConfig = RequestConfig.custom()
                .setSocketTimeout(10000)
                .setConnectTimeout(10000)
                .setConnectionRequestTimeout(10000)
                .build();
        //配置HttpClient
        HttpClient httpClient = HttpClients.custom()
                .setDefaultRequestConfig(requestConfig)
                .build();
        HttpGet httpGet = new HttpGet("http://www.********.com.cn/b.asp");
        HttpResponse response = null;
```

```java
try {
    response = httpClient.execute(httpGet);   //执行请求
}catch (Exception e){
    e.printStackTrace();
}
String result = EntityUtils.toString(response.getEntity(),
"gbk");                                        //获取HTML格式的结果
System.out.println(result);                    //输出结果
```

另外，也可针对实例化的请求方法设置超时时间，如程序 4-35 所示。

程序 4-35

```java
//实例化 HttpClient
HttpClient httpClient = HttpClients.createDefault();
HttpGet httpGet=new HttpGet("http://www.********.com.cn/b.asp");//GET请求
RequestConfig requestConfig = RequestConfig.custom()
        .setSocketTimeout(2000)
        .setConnectTimeout(2000)
        .setConnectionRequestTimeout(10000)
        .build();//设置超时时间
httpGet.setConfig(requestConfig);              //请求方法配置信息
HttpResponse response = null;
try {
    response = httpClient.execute(httpGet); //执行请求
}catch (Exception e){
    e.printStackTrace();
}
String result = EntityUtils.toString(response.getEntity(),
"gbk");                                        //获取HTML格式的结果
System.out.println(result);                    //输出结果
```

4.2.7 代理服务器的使用

在 4.2.6 节中，已经介绍了 RequestConfig 类，这里就直接使用该类配置代理服务器访问指定 URL，如程序 4-36 所示。

程序 4-36

```java
RequestConfig defaultRequestConfig = RequestConfig.custom()
        .setProxy(new HttpHost("171.221.239.11",808, null))
        .build();    //添加代理
```

```
            HttpGet httpGet = new HttpGet("http://www.********.com.cn/b.asp");
            HttpClient httpClient = HttpClients.custom().
                    setDefaultRequestConfig(defaultRequestConfig).build();  //配置HttpClient
            //执行请求
            HttpResponse httpResponse = httpClient.execute(httpGet);
            if (httpResponse.getStatusLine().getStatusCode() == 200){
                String result = EntityUtils.toString(httpResponse.getEntity(),"gbk");
                System.out.println(result);   //输出结果
            }
```

另外,与设置超时时间相同,也可针对实例化的请求方法配置代理服务器,如程序 4-37 所示。

程序 4-37

```
            HttpClient httpClient = HttpClients.custom()
                    .build();  //实例化HttpClient
            // 配置代理
            HttpHost proxy = new HttpHost("171.221.239.11",808, null);
            RequestConfig config = RequestConfig.custom().setProxy(proxy).build();
            HttpGet httpGet = new HttpGet("http://www.********.com.cn/b.asp");
            httpGet.setConfig(config);  //针对实例化的请求方法配置代理
            HttpResponse httpResponse = httpClient.execute(httpGet);
            if (httpResponse.getStatusLine().getStatusCode() == 200){
                String result = EntityUtils.toString(httpResponse.getEntity(),"gbk");
                System.out.println(result);
            }
```

4.2.8 文件下载

下载 HTML、图片、PDF 和压缩等文件时,一种方法是使用 HttpEntity 类将响应实体转化成字节数组,再利用输出流的方式写入指定文件。另一种方法是使用 HttpEntity 类中的 writeTo(OutputStream)方法,直接将响应实体写入指定的输出流中,这种方法简单且常用,如程序 4-38,使用 writeTo(OutputStream)下载 ".tar.gz" 格式的压缩文件。图 4.18 所示为程序 4-38 下载的文件。

程序 4-38

```java
import java.io.FileOutputStream;
import java.io.IOException;
import java.io.OutputStream;
import org.apache.http.HttpResponse;
import org.apache.http.client.HttpClient;
import org.apache.http.client.methods.HttpGet;
import org.apache.http.impl.client.HttpClients;
import org.apache.http.util.EntityUtils;
public class HttpclientDownloadFile {
    public static void main(String[] args) throws IOException {
        String url = "https://www-us.******.org/dist//httpd/httpd-2.4.37.tar.gz";
        //初始化HttpClient
        HttpClient httpClient = HttpClients.custom().build();
        HttpGet httpGet = new HttpGet(url);
        //获取结果
        HttpResponse httpResponse = null;
        try {
            httpResponse = httpClient.execute(httpGet);
        } catch (IOException e) {
            e.printStackTrace();
        }
        //非常简单的下载文件的方法
        OutputStream out = new FileOutputStream("file/httpd-2.4.37.tar.gz");
        httpResponse.getEntity().writeTo(out);
        EntityUtils.consume(httpResponse.getEntity()); //消耗实体
    }
}
```

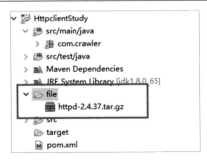

图 4.18　执行程序 4-38 下载的文件

4.2.9 HTTPS 请求认证

与 Jsoup 类似，使用 HttpClient 直接请求以 https://为前缀的 URL，有时也会产生图 4.11 所示的错误信息，即找不到合法证书请求目标 URL。如使用程序 4-27 请求下面的 URL。

```
https://cn.*******.com
```

程序 4-39 提供了一种解决方案。首先，利用内部类 SSL509TrustManager，创建 X.509 证书信任管理器；之后，使用 SSLConnectionSocketFactory()方法创建 SSL 连接，并利用 Registry 注册 http 和 https 套接字工厂；接着，使用 PoolingHttpClientConnectionManager()方法实例化连接池管理器；最后，基于实例化的连接池管理器和 RequestConfig 配置的信息，来实例化一个可以执行 HTTPS 请求的 HttpClient。

在程序 4-40 中，调用程序 4-39 的 SSLClient 类中的 HttpClient initSSLClient(String SSLProtocolVersion)方法，其中，参数 SSLProtocolVersion 设置为"SSLv3"，基于该方法实例化的 HttpClient 能够成功请求以 https://为前缀的 URL。

程序 4-39

```java
import java.security.KeyManagementException;
import java.security.NoSuchAlgorithmException;
import java.security.cert.X509Certificate;
import java.util.Arrays;
import javax.net.ssl.SSLContext;
import javax.net.ssl.X509TrustManager;
import org.apache.http.client.HttpClient;
import org.apache.http.client.config.AuthSchemes;
import org.apache.http.client.config.CookieSpecs;
import org.apache.http.client.config.RequestConfig;
import org.apache.http.config.Registry;
import org.apache.http.config.RegistryBuilder;
import org.apache.http.conn.socket.ConnectionSocketFactory;
import org.apache.http.conn.socket.PlainConnectionSocketFactory;
import org.apache.http.conn.ssl.NoopHostnameVerifier;
import org.apache.http.conn.ssl.SSLConnectionSocketFactory;
import org.apache.http.impl.client.HttpClients;
import org.apache.http.impl.conn.PoolingHttpClientConnectionManager;
public class SSLClient {
```

```java
/**
 * 基于SSL配置HttpClient
 * @param  SSLProtocolVersion(SSL, SSLv3, TLS, TLSv1, TLSv1.1, TLSv1.2)
 * @return HttpClient
 */
public HttpClient initSSLClient(String SSLProtocolVersion){
    RequestConfig defaultConfig = null;
    PoolingHttpClientConnectionManager pcm = null;
    try {
        X509TrustManager xtm = new SSL509TrustManager(); //创建信任管理
        //创建SSLContext对象，并使用指定的信任管理器初始化
        SSLContext context = SSLContext.getInstance(SSLProtocolVersion);
        context.init(null, new X509TrustManager[]{xtm}, null);
        /*从SSLContext对象中得到SSLConnectionSocketFactory对象
        *NoopHostnameVerifier.INSTANCE表示接受任何有效的和符合目标主机的SSL会话
        */
        SSLConnectionSocketFactory sslConnectionSocketFactory =
                new SSLConnectionSocketFactory(context, NoopHostnameVerifier.INSTANCE);
        //设置全局请求配置，包括cookie规范
        defaultConfig = RequestConfig.custom().setCookieSpec(CookieSpecs.STANDARD_STRICT)
                .setExpectContinueEnabled(true)
                .setTargetPreferredAuthSchemes(Arrays.asList(AuthSchemes.NTLM, AuthSchemes.DIGEST))
                .setProxyPreferredAuthSchemes(Arrays.asList(AuthSchemes.BASIC)).build();
        // 注册http和https套接字工厂
        Registry<ConnectionSocketFactory> sfr = RegistryBuilder.<ConnectionSocketFactory>create()
                .register("http", PlainConnectionSocketFactory.INSTANCE)
                .register("https", sslConnectionSocketFactory).build();
        //基于sfr创建连接管理器
        pcm = new PoolingHttpClientConnectionManager(sfr);
    }catch(NoSuchAlgorithmException | KeyManagementException e){
        e.printStackTrace();
    }
```

```
            //基于连接管理器和配置,实例化HttpClient
            HttpClient httpClient = HttpClients.custom().
setConnectionManager(pcm).setDefaultRequestConfig(defaultConfig)
                    .build();
            return httpClient;
    }
    //实现X509TrustManager接口
    private static class SSL509TrustManager implements
X509TrustManager {
        //检查客户端证书
        public void checkClientTrusted(X509Certificate[]
x509Certificates, String s) {
            //do nothing 接受任意客户端证书
        }
        //检查服务器端证书
        public void checkServerTrusted(X509Certificate[]
x509Certificates, String s) {
            //do nothing 接受任意服务器端证书
        }
        //返回受信任的X509证书
        public X509Certificate[] getAcceptedIssuers() {
            return new X509Certificate[0];
        }
    };
}
```

程序4-40

```
    import org.apache.http.HttpResponse;
    import org.apache.http.HttpStatus;
    import org.apache.http.ParseException;
    import org.apache.http.client.HttpClient;
    import org.apache.http.client.methods.HttpGet;
    import org.apache.http.util.EntityUtils;
    public class Test {
      public static void main(String[] args) throws ParseException,
IOException {
            String url = "https://cn.*******.com";
            SSLClient sslClient = new SSLClient();    //实例化
            HttpClient httpClientSSL = sslClient.initSSLClient("SSLv3");
            HttpGet httpGet = new HttpGet(url);
            //获取结果
```

```java
        HttpResponse httpResponse = null;
        try {
            httpResponse = httpClientSSL.execute(httpGet);
        } catch (IOException e) {
            e.printStackTrace();
        }
        if(httpResponse.getStatusLine().getStatusCode() == HttpStatus.SC_OK){  //状态码200表示响应成功
            //获取实体内容
            String entity = EntityUtils.toString(httpResponse.getEntity(),"UTF-8");
            //输出实体内容
            System.out.println(entity);
            EntityUtils.consume(httpResponse.getEntity());   //消耗实体
        }else {
            //关闭HttpEntity的流实体
            EntityUtils.consume(httpResponse.getEntity());   //消耗实体
        }
    }
}
```

4.2.10 请求重试

使用 HttpClient 请求 URL 时，有时会出现请求异常的情况。针对一些非致命的异常，可以通过请求重试解决。HttpClient 提供了默认重试策略 DefaultHttpRequestRetryHandler。DefaultHttpRequestRetryHandler 类实现了 HttpRequestRetryHandler 接口，重写了 retryRequest()方法。程序 4-41 为 DefaultHttpRequestRetryHandler 类的一部分源码。由源码可以发现 DefaultHttpRequestRetryHandler 类定义的默认重试次数为 3 次；幂等方法（如 GET 和 HEAD 是幂等的）可以重试；如果网页请求失败，可以重试。另外，针对 4 种异常不进行重试，这四种异常分别是 InterruptedIOException（线程中断异常）、UnknownHostException（未知的 Host 异常）、ConnectException（连接异常，如连接拒绝异常）和 SSLException（HTTPS 请求认证异常）。

程序 4-41

```java
/** 重试次数 */
private final int retryCount;
/** 如果请求发送成功，是否还会被再次请求 */
private final boolean requestSentRetryEnabled;
```

```java
    private final Set<Class<? extends IOException>> nonRetriableClasses;
    /** 发生InterruptedIOException、UnknownHostException、
ConnectException、SSLException错误不重试 */
    protected DefaultHttpRequestRetryHandler(
            final int retryCount,
            final boolean requestSentRetryEnabled,
            final Collection<Class<? extends IOException>> clazzes) {
        super();
        this.retryCount = retryCount;
        this.requestSentRetryEnabled = requestSentRetryEnabled;
        this.nonRetriableClasses = new HashSet<Class<? extends IOException>>();
        for (final Class<? extends IOException> clazz: clazzes) {
            this.nonRetriableClasses.add(clazz);
        }
    }
    /** 自由设置请求次数 */
    public DefaultHttpRequestRetryHandler(final int retryCount, final boolean requestSentRetryEnabled) {
        this(retryCount, requestSentRetryEnabled, Arrays.asList(
                InterruptedIOException.class,
                UnknownHostException.class,
                ConnectException.class,
                SSLException.class));
    }
    /** 设置了默认重试次数 */
    public DefaultHttpRequestRetryHandler() {
        this(3, false);
    }
    /** 重写了RetryRequest方法 */
    @Override
    public boolean retryRequest(
            final IOException exception,
            final int executionCount,
            final HttpContext context) {
        Args.notNull(exception, "Exception parameter");
        Args.notNull(context, "HTTP context");
        /**如果重试次数大于设置的重试次数，则停止重试*/
        if (executionCount > this.retryCount) {
            return false;
        }
        /**如果是上面的4种异常，则停止重试*/
```

```
            if (this.nonRetriableClasses.contains(exception.getClass())) {
                return false;
            } else {
                for (final Class<? extends IOException> rejectException :
this.nonRetriableClasses) {
                    if (rejectException.isInstance(exception)) {
                        return false;
                    }
                }
            }
            final HttpClientContext clientContext = HttpClientContext.
adapt(context);
            final HttpRequest request = clientContext.getRequest();
        /**请求是否终止,如果终止停止重试*/
            if(requestIsAborted(request)){
                return false;
            }
        /**请求是否是幂等方法,如果是则重试*/
            if (handleAsIdempotent(request)) {
                return true;
            }
        /** clientContext.isRequestSent()判断请求是否发送成功,没成功则重试*/
            if (!clientContext.isRequestSent() || this.
requestSentRetryEnabled) {
                return true;
            }
        /**其他情况*/
            return false;
        }
```

在实例化 HttpClient 时,可以使用 HttpClientBuilder 类中的 setRetryHandler()方法设置重试,如程序 4-42 提供了两种方式。其中,第一种使用默认重试次数 3 次,第二种自定义重试次数为 5 次。

程序 4-42

```
HttpClient httpClient = HttpClients.custom()
    .setRetryHandler(new DefaultHttpRequestRetryHandler())
    .build();
//自定义重试次数
HttpClient httpClient = HttpClients.custom()
    .setDefaultRequestConfig(defaultConfig)
```

```
            .setRetryHandler(new DefaultHttpRequestRetryHandler(5, true))
            .build();
```

值得注意的是，在进行数据爬取时经常遇到的两种超时时间：ConnectTimeout（建立连接的超时时间）和 SocketTimeout（获取数据的超时时间），这两种超时时间对应的异常（ConnectTimeoutException 与 SocketTimeoutException）都继承自 InterruptedIOException 类，即属于线程中断异常，不会进行重试。

4.2.11 多线程执行请求

4.5 版本的 HttpClient 中的连接池管理器 PoolingHttpClientConnectionManager 类实现了 HTTP 连接池化管理，其管理连接的单位为路由（Route），每个路由维护一定数量（默认是 2）的连接；当给定路由的所有连接都被租用时，则新的连接请求将发生阻塞，直到某连接被释放回连接池。另外，PoolingHttpClientConnectionManager 维护的连接次数也受总数 MaxTotal（默认是 20）的限制。

当 HttpClient 配置了 PoolingHttpClientConnectionManager 时，其可以同时执行多个 HTTP 请求，即实现多线程操作。程序 4-43 提供了一个简单的多线程请求多个 URL 案例。由程序 4-43 可知，使用实例化的 PoolingHttpClientConnectionManager 可以设置最大连接数、每个路由的最大连接数、Connection 信息和 Socket 信息等。另外，本案例是通过继承 Thread 类，重写 Thread 类的 run() 方法实现的多线程，有兴趣的读者，也可通过实现 Runnable 接口的方式实现多线程。图 4.19 所示为程序 4-43 在控制台输出的结果，图 4.20 所示为程序 4-43 下载的 HTML 文件。

程序 4-43

```
public class Test {
    public static void main(String[] args) throws FileNotFoundException {
        //添加连接参数
        ConnectionConfig connectionConfig = ConnectionConfig.custom()
                .setMalformedInputAction(CodingErrorAction.IGNORE)
                .setUnmappableInputAction(CodingErrorAction.IGNORE)
                .setCharset(Consts.UTF_8)
                .build();
        //添加socket参数
        SocketConfig socketConfig = SocketConfig.custom()
                .setTcpNoDelay(true)
                .build();
        //配置连接池管理器
```

```java
        PoolingHttpClientConnectionManager pcm = new PoolingHttpClientConnectionManager();
        // 设置最大连接数
        pcm.setMaxTotal(100);
        // 设置每个连接的路由数
        pcm.setDefaultMaxPerRoute(10);
        //设置连接信息
        pcm.setDefaultConnectionConfig(connectionConfig);
        //设置socket信息
        pcm.setDefaultSocketConfig(socketConfig);
        //设置全局请求配置，包括cookie规范、Http认证、超时时间
        RequestConfig defaultConfig = RequestConfig.custom()
                .setCookieSpec(CookieSpecs.STANDARD_STRICT)
                .setExpectContinueEnabled(true)
                .setTargetPreferredAuthSchemes(Arrays
                        .asList(AuthSchemes.NTLM, AuthSchemes.DIGEST))
                .setProxyPreferredAuthSchemes(Arrays.asList(AuthSchemes.BASIC))
                .setConnectionRequestTimeout(30*1000)
                .setConnectTimeout(30*1000)
                .setSocketTimeout(30*1000)
                .build();
        CloseableHttpClient httpClient = HttpClients.custom()
                .setConnectionManager(pcm)
                .setDefaultRequestConfig(defaultConfig)
                .build();
        // 请求的URL
        String[] urlArr = {
                "http://www.********.com.cn/html/index.asp",
                "http://www.********.com.cn/html/html_basic.asp",
                "http://www.********.com.cn/html/html_elements.asp",
                "http://www.********.com.cn/html/html_attributes.asp",
                "http://www.********.com.cn/html/html_formatting.asp"
        };
        //创建固定大小的线程池
        ExecutorService exec = Executors.newFixedThreadPool(3);
        for(int i = 0; i< urlArr.length;i++){
            //HTML需要输出的文件名
            String filename = urlArr[i].split("html/")[1];
            //创建HTML文件输出目录
```

```java
                OutputStream out = new FileOutputStream("file/" + filename);
                HttpGet httpget = new HttpGet(urlArr[i]);
                //启动线程执行请求
                exec.execute(new DownHtmlFileThread(httpClient, httpget, out));
            }
            //关闭线程
            exec.shutdown();
        }
        static class DownHtmlFileThread extends Thread {
            private final CloseableHttpClient httpClient;
            private final HttpContext context;
            private final HttpGet httpget;
            private final OutputStream out;
            //输入的参数
            public DownHtmlFileThread(CloseableHttpClient httpClient,
                    HttpGet httpget, OutputStream out) {
                this.httpClient = httpClient;
                this.context = HttpClientContext.create();
                this.httpget = httpget;
                this.out = out;
            }
            @Override
            public void run() {
                System.out.println(Thread.currentThread().getName()
                        + "线程请求的URL为:" + httpget.getURI());
                try {
                    CloseableHttpResponse response = httpClient.execute(
                            httpget, context);   //执行请求
                    try {
                        //将HTML文档写入文件
                        out.write(EntityUtils.toString(response.getEntity(),"gbk")
                                .getBytes());
                        out.close();
                        //消耗实体
                        EntityUtils.consume(response.getEntity());
                    } finally{
                        response.close();   //关闭响应
                    }
```

```
        } catch (ClientProtocolException ex) {
            ex.printStackTrace(); // 处理 Protocol错误
        } catch (IOException ex) {
            ex.printStackTrace(); // 处理I/O错误
        }
    }
  }
}
```

```
pool-1-thread-1线程请求的URL为:http://www.▇▇▇.com.cn/html/index.asp
pool-1-thread-2线程请求的URL为:http://www.▇▇▇.com.cn/html/html_basic.asp
pool-1-thread-3线程请求的URL为:http://www.▇▇▇.com.cn/html/html_elements.asp
pool-1-thread-2线程请求的URL为:http://www.▇▇▇.com.cn/html/html_attributes.asp
pool-1-thread-3线程请求的URL为:http://www.▇▇▇.com.cn/html/html_formatting.asp
```

图 4.19　程序 4-43 在控制台输出的结果

图 4.20　程序 4-43 下载的 HTML 文件

4.3　URLConnection 与 HttpURLConnection

URLConnection 是 java.net 包中的一个抽象类，其主要用于实现应用程序与 URL 之间的通信。HttpURLConnection 继承自 URLConnection，也是抽象类。在网络爬虫中，可以使用 URLConnection 或 HttpURLConnection 请求 URL 获取流数据，通过对流数据的操作，获取具体的实体内容。

4.3.1　实例化

URLConnection 与 HttPURLConnection 都是抽象类，无法直接创建实例化对象，但可以通过 java.net 包 URL 类中的 openConnection()方法创建 URLConnection 与 HttPURLConnection 实例。程序 4-44 分别给出了实例化 URLConnection 和 HttpURLConnection 的方式。

程序 4-44

```
        URL url = new URL("http://www.********.com.cn/b.asp");
        URLConnection conn = url.openConnection();
        HttpURLConnection conn = (HttpURLConnection) url.openConnection();
```

4.3.2 获取网页内容

要获取 URLConnection 请求到的实体内容，需通过数据流操作。在 openConnection()方法执行完毕后，通过 getInputStream()方法获取输入流，之后采用 BufferedReader 读取输入流信息，操作如程序 4-45 所示。基于该程序，可成功获取指定 URL 对应的网页数据。

程序 4-45

```
        URL url = new URL("http://www.********.com.cn/b.asp");
        //URLConnection conn = url.openConnection();
        HttpURLConnection conn = (HttpURLConnection) url.openConnection();
        //获取流数据
        InputStream in=conn.getInputStream();
        // 定义BufferedReader输入流来读取响应实体内容
        BufferedReader reader = new BufferedReader(
            new InputStreamReader(in));
        String line;
        String html = "";
        while ((line = reader.readLine()) != null) {
            html += line;
        }
        System.out.println(html);
        reader.close();
```

4.3.3 GET 请求

针对实例化的 HttpURLConnection，可以使用 setRequestMethod(String method)方法设置 HTTP 请求方法，其可设置的请求方法包括 GET、POST、HEAD、OPTIONS、PUT、DELETE 以及 TRACE。程序 4-46 演示了设置 GET 的操作。在程序 4-46 中，setDoInput(true)表示 URL 连接可用于输入，setRequestMethod("GET")表示设置的请求

方法为 GET。基于 getResponseCode()方法可以获取响应状态码,如果该状态码为 200,则利用实例化的 StringBuffer 将响应内容读取出来。

程序 4-46

```java
        //初始化URL
        URL url = new URL("http://www.********.com.cn/b.asp");
        HttpURLConnection conn = (HttpURLConnection) url.openConnection();
        //允许Input
        conn.setDoInput(true);
        conn.setRequestMethod("GET");              //设置请求的方法
        conn.connect();                            //连接操作
        int statusCode = conn.getResponseCode();   //获取响应状态码
        String responseBody = null;
        //如果响应状态码为200
        if (HttpURLConnection.HTTP_OK == statusCode) {
            // 定义BufferedReader输入流来读取URL的响应,这里设置编码
            BufferedReader bufferedReader = new BufferedReader(
                new InputStreamReader(conn.getInputStream(), "GBK"));
            //读取内容
            String readLine = null;
            StringBuffer response = new StringBuffer();
            while (null != (readLine = bufferedReader.readLine())) {
                response.append(readLine);
            }
            bufferedReader.close();
            responseBody = response.toString();
        }
        System.out.println(responseBody);
```

4.3.4 模拟提交表单(POST 请求)

使用 HttpURLConnection 也可以提交 POST 请求。以下仍以 4.1.4 节中的某快递网站提交表单为例,具体请求的地址、查询的快递单号以及抓包获得的参数如下所示。

请求地址:http://www.*****.com/ems.php。

快递单号:EH629625211CS。

抓包获得的参数:wen:EH629625211CS action:ajax rnd: 0.15938420328106995。

操作程序如 4-47 所示。在使用 POST 提交参数时，必须将 setDoOutput(boolean dooutput)方法中的参数设置为 true。运行程序 4-47 能够成功获取服务器返回的数据，如图 4.21 所示。

程序 4-47

```java
//POST 表单需要提交的参数
String wen = "EH629625211CS";
String action = "ajax";
//初始化URL
URL url = new URL("http://www.*****.com/ems.php");
HttpURLConnection conn = (HttpURLConnection) url.openConnection();
//允许Output
conn.setDoOutput(true);
conn.setRequestMethod("POST");         //POST提交参数
StringBuffer params = new StringBuffer();
// 表单参数拼接
params.append("wen").append("=").append(wen).append("&")
.append("action").append("=").append(action);
byte[] bypes = params.toString().getBytes();
conn.getOutputStream().write(bypes);//在连接中添加参数
// 定义BufferedReader输入流来读取URL的响应，这里设置编码
BufferedReader bufferedReader = new BufferedReader(
        new InputStreamReader(conn.getInputStream(),
"utf-8"));
String line;
String html = "";
while ((line = bufferedReader.readLine()) != null) {
    html += line;
}
System.out.println(html);
bufferedReader.close();
```

`<span style="color:red;">EMS快递单号【EH629625211CS】暂无信息,可尝试刷新一遍试试！！</span>`

图 4.21　程序 4-47 的输出结果

4.3.5　设置头信息

针对初始化的 URLConnection 及 HttpURLConnection，可以使用 setRequestProperty(key,value)方法设置具体的请求头信息，如程序 4-48 所示。

程序 4-48

```java
    //初始化 URL
    URL url = new URL("http://www.********.com.cn/b.asp");
    URLConnection conn = url.openConnection();
    //HttpURLConnection conn = (HttpURLConnection) url.openConnection()
    //添加请求头信息
    conn.setRequestProperty("Accept", "text/html");
    conn.setRequestProperty("Accept-Language", "zh-CN,zh;q=0.9");
    conn.setRequestProperty("Host", "www.********.com.cn");
    conn.setRequestProperty("Cache-Control", "max-age=0");
    conn.setRequestProperty("User-Agent", "Mozilla/5.0 (Windows NT 10.0; Win64; x64) AppleWebKit/537.36 (KHTML, like Gecko) Chrome/63.0.3239.108 Safari/537.36");
    conn.connect();
    BufferedReader bufferedReader = new BufferedReader(
            new InputStreamReader(conn.getInputStream(), "gbk"));
    String line;
    String html = "";
    while ((line = bufferedReader.readLine()) != null) {
        html += line;
    }
    System.out.println(html);
    bufferedReader.close();
}
```

4.3.6 连接超时设置

使用 URLConnection 与 HttpURLConnection 时，可以设置两种超时时间，分别是连接超时时间（ConnectTimeout）和读取超时时间（ReadTimeout），如程序 4-49 所示。

程序 4-49

```java
    URL url = new URL("http://www.********.com.cn/b.asp");
    URLConnection conn = url.openConnection();
    // HttpURLConnection conn = (HttpURLConnection) url.openConnection()
    conn.setConnectTimeout(30000);    //连接超时，单位为毫秒
    conn.setReadTimeout(30000);       //读取超时，单位为毫秒
```

4.3.7 代理服务器的使用

针对 URLConnection 与 HttpURLConnection，可以使用 Proxy 设置代理，如程序 4-50 所示。

程序 4-50

```
//代理的 IP 及端口设置
Proxy proxy = new Proxy(Proxy.Type.HTTP, new InetSocketAddress
("171.97.67.160", 3128));
URL url = new URL("http://www.********.com.cn/b.asp");
URLConnection conn = url.openConnection(proxy);     //添加代理
conn.connect();                                      //建立连接
```

4.3.8 HTTPS 请求认证

使用 URLConnection 与 HttpURLConnection 直接访问一些以 https:// 为前缀的 URL 时，也会产生如图 4.11 所示的错误。为此，在使用 URLConnection 与 HttpURLConnection 之前，也需要创建信任管理器（忽略证书验证）。程序 4-51 所示为具体的实现方式，可以发现这里的 initUnSecureTSL() 方法与 4.1.8 节创建的 initUnSecureTSL() 方法内容完全一致。

程序 4-51

```java
import java.io.BufferedReader;
import java.io.IOException;
import java.io.InputStreamReader;
import java.net.HttpURLConnection;
import java.net.URL;
import java.security.cert.X509Certificate;
import javax.net.ssl.HttpsURLConnection;
import javax.net.ssl.SSLContext;
import javax.net.ssl.SSLSocketFactory;
import javax.net.ssl.TrustManager;
import javax.net.ssl.X509TrustManager;
public class URLConnectionSSL {
    public static void main(String[] args) throws IOException {
        initUnSecureTSL();
        //使用URLConnection请求数据
        URL url = new URL("https://cn.*******.com");
```

```java
            HttpURLConnection conn = (HttpURLConnection) url.openConnection();
                int statusCode = conn.getResponseCode(); //获取响应状态码
            String responseBody = null;
            //如果响应状态码为200
            if (HttpURLConnection.HTTP_OK == statusCode) {
                // 定义BufferedReader输入流来读取URL的响应，这里设置编码
                BufferedReader bufferedReader = new BufferedReader(
                        new InputStreamReader(conn.getInputStream(), "utf-8"));
                //读取内容
                String readLine = null;
                StringBuffer response = new StringBuffer();
                while (null != (readLine = bufferedReader.readLine())) {
                    response.append(readLine);
                }

                bufferedReader.close();
                responseBody = response.toString();
            }
            System.out.println(responseBody);
    }
    private static void initUnSecureTSL() {
        // 创建信任管理器(不验证证书)
        final TrustManager[] trustAllCerts = new TrustManager[]{new X509TrustManager() {
            //检查客户端证书
            public void checkClientTrusted(final X509Certificate[] chain, final String authType) {
                //do nothing 接受任意客户端证书
            }
            //检查服务器端证书
            public void checkServerTrusted(final X509Certificate[] chain, final String authType) {
                //do nothing 接受任意服务器端证书
            }
            //返回受信任的X509证书
            public X509Certificate[] getAcceptedIssuers() {
                return null; //或者return new X509Certificate[0];
            }
        }};
        try {
```

```
            // 创建SSLContext对象，并使用指定的信任管理器初始化
            SSLContext sslContext = SSLContext.getInstance("SSL");
            sslContext.init(null, trustAllCerts, new java.security.SecureRandom());
            //基于信任管理器创建套接字工厂
            SSLSocketFactory sslSocketFactory = sslContext.getSocketFactory();
            //为HttpsURLConnection配置套接字工厂
            HttpsURLConnection.setDefaultSSLSocketFactory(sslSocketFactory);
            //正常访问Https协议网站
        } catch (Exception e) {
            e.printStackTrace();
        }
    }
}
```

4.4 本章小结

本章主要介绍了 Jsoup、HttpClient、URLConnection 与 HttpURLConnection 在网页内容获取方面的应用。通过示例程序演示了这几种工具的使用方式以及需要注意的细节。在实际应用中，读者可根据自己的熟练程度选择使用的工具。

第5章 网页内容解析

5.1 HTML 解析

5.1.1 CSS 选择器

CSS（Cascading Style Sheets），即层叠样式表，主要用于 HTML 文档的样式化与布局，具体涉及字体、颜色、编辑和高级定位等。CSS Selector，即 CSS 选择器，是用于匹配元素（Elements）的一种模式。在网络爬虫中，常使用 CSS 选择器，定位 HTML 文档中的元素，进而抽取 HTML 文档中的相应字段。下面将分别介绍 4 类 CSS 选择器。

1. 基础选择器

基础选择器包括类别选择器、标签选择器、id 选择器和通用元素选择器，如表 5.1 所示。

表 5.1　基础选择器

选择器	描述	案例	含义
.class	类别选择器	.intro	匹配 class="intro"的所有元素
element	标签选择器	p	匹配标签为<p>的元素
#id	id 选择器	#navsecond	选择 id="navsecond "的所有元素
*	通用元素选择器	*	选择所有元素

2. 属性选择器

属性选择器允许用户自定义属性名称，而不仅限于基础选择器中的 id 和 class 属性，如表 5.2 所示。

表 5.2　属性选择器

选择器	含义
[attribute]	选择带有 title 属性的所有元素
[^attribute]	利用属性前缀查找元素。例如，查找属性前缀为 titl 的所有元素
[attribute="value"]	选择 title="CSS 教程"的所有元素
[attribute~="value"]	选择 title 属性包含"CSS"的所有元素，这种方式也用于正则匹配

续表

选择器	含义
[attribute^="value"]	选择 title 属性最前面为"CSS"的所有元素
[attribute$="value"]	选择 title 属性最后面为"CSS"的所有元素

另外，属性选择器，也可以指定具体标签的属性。使用方式是在方括号前加上标签名称，案例如下。

```
div[attribute]
div[^attribute]
```

3．组合选择器

为了匹配特定的元素，有时需要组合使用不同的选择器。组合选择器包括多元素选择器、后代选择器、子代选择器、兄弟选择器、直接相邻兄弟选择器和任意组合选择器，如表 5.3 所示。

表 5.3 组合选择器

选择器	描述	案例	含义
element,element	多元素选择器（以逗号分隔）	div,p	选择所有 \<div\> 元素和所有 \<p\> 元素
ancestor child	后代选择器（查找某个元素下的子元素，用空格分隔）	div p	匹配\<div\>元素内部的所有\<p\>元素
parent > child	子代选择器（查找某个父元素下的直接子元素）	div.content > p	查找 div[class="content"]下一层的\<p\>元素
siblingA ~ siblingX	兄弟选择器	div~p	选择\<div\>元素之后的所有同级\<p\>元素
siblingA + siblingB	直接相邻兄弟选择器	div+p	选择\<div\>元素之后的下一个同级\<p\>元素

4．伪选择器

伪选择器，以冒号为前缀。表 5.4 列举了一些伪选择器。

表 5.4 伪选择器

选择器	描述	案例	含义
:lt(n)	查找同级索引值小于 n 的元素	td:lt(3)	小于 3 列的元素
:gt(n)	查找同级索引值大于 n 的元素	div p:gt(2)	\<div\>中有包含 2 个以上的\<p\>元素
:eq(n)	查找同级索引值与 n 相等的元素	form input:eq(1)	包含 1 个 input 标签的\<form\>元素

续表

选 择 器	描 述	案 例	含 义
:has(selector)	匹配包含某元素的元素	div:has(p)	查找包含\<p\>元素的\<div\>元素
:not(selector)	查找与选择器不匹配的元素	div:has(p)	不包含\<p\>元素的所有\<div\>元素
:contains(text)	查找包含指定文本的元素	p:contains(java)	包含 java 文本的所有\<p\>元素
:nth-child(n)	查找某父元素下的第 n 个子元素	div:nth-child(5)	某父元素下的 5 个\<div\>元素

5.1.2 Xpath 语法

在网络爬虫中，常用 Xpath 语法定位所要解析的内容。Xpath 语法使用路径表达式来选取 HTML 或 XML 文档中的节点或节点集合。其中，节点可以是元素、属性、注释和文本等内容，Xpath 语法如表 5.5 所示。

表 5.5　Xpath 语法

表 达 式	描 述	案 例	案 例 含 义
nodename	选取此节点的所有子节点	body	选取\<body\>元素的所有子节点
/	从根节点选取	/html	选取根节点\<html\>
//	从选择的当前节点选择文档中的节点，而不考虑它们的位置	//div	选取所有\<div\>节点，而不管它们在文档的位置
.	选取当前节点	./p	选取当前的\<p\>节点
@	选取属性	//a[@href]	选取所有拥有属性 href 的\<a\>节点
@	选取属性	//div[@id='course']	选取所有 id 属性为 course 的\<div\>节点

在使用 Xpath 语法解析 HTML 文件时，通常需要组合使用表 5-5 所示语法。表 5.6 所示为 Xpath 语法的组合使用。

表 5.6　Xpath 语法的组合使用

案 例	案 例 含 义
//div[@id='w3school']/h1	选取所有 id 属性为 w3school 的\<div\>节点下的\<h1\>节点
//body/a[1]	选取\<body\>下的第一个\<a\>节点
//body//a[last()]	选取\<body\>下的最后一个\<a\>节点

5.1.3 Jsoup 解析 HTML

第 4 章的 4.1 节介绍了如何使用 Jsoup 获取网页数据，本节将重点介绍 Jsoup 的解析功能。Jsoup 依赖于 CSS 选择器与 jQuery 方法来操作 HTML 文件中的数据。使用 Jsoup 之前，需要了解 Jsoup 中的 Node、Element 和 Document 的概念。

节点（Node）：HTML 文件中所包含的内容都可以看成一个节点。节点有很多种类型，如属性（Attribute）节点、注释（Note）节点、文本（Text）节点和元素（Element）节点等。解析 HTML 文件的过程，实际上就是对节点进行操作的过程。

元素（Element）：节点的子集，所以一个元素也是一个节点。

文档（Document）：整个 HTML 文档的源码内容。

下面介绍 Jsoup 解析 HTML 文件的方法。

1. 解析静态 HTML 文件

给定 HTML 字符串，可以使用 org.jsoup.Jsoup 类中的 parse(String html) 方法，将 String 类型的 HTML 文件转化成 Document 类型。之后，可以使用 org.jsoup.nodes.Element 类中的 select(String cssQuery) 方法定位到所要解析的元素。程序 5-1 所示为一个简单的案例，图 5.1 所示为执行程序 5-1 所输出的结果。

程序 5-1

```java
import java.io.IOException;
import org.jsoup.Jsoup;
import org.jsoup.nodes.Document;
import org.jsoup.nodes.Element;
public class JsoupParseStaticFile {
 public static void main(String[] args) throws IOException {
    //HTML静态文件
    String html = "<html><body><div id=\"w3school\"> <h1>浏览器脚本教程</h1> <p><strong>从左侧的菜单选择你需要的教程！</strong></p></div>"
            + "<div> <div id=\"course\"> <ul> <li><a href=\"/js/index.asp\" title=\"JavaScript 教程\">JavaScript</a></li> </ul> </div> </body></html>";
      //转化成Document
      Document doc = Jsoup.parse(html);
      //基于CSS选择器获取元素，也可写成[id=w3school]
      Element element = doc.select("div[id=w3school]").get(0);
      System.out.println("输出解析的元素内容为:");
      System.out.println(element);
```

```
        //从Element提取内容(抽取一条Node对应的信息)
        String text1 = element.select("h1").text();
        //从Element提取内容(抽取一条Node对应的信息)
        String text2 = element.select("p").text();
        System.out.println("抽取的文本信息为:");
        System.out.println(text1 + "\t" + text2);
    }
}
```

```
输出解析的元素内容为:
<div id="w3school">
 <h1>浏览器脚本教程</h1>
 <p><strong>从左侧的菜单选择你需要的教程!</strong></p>
</div>
抽取的文本信息为:
浏览器脚本教程          从左侧的菜单选择你需要的教程!
```

图 5.1　执行程序 5-1 所输出的结果

2. 解析 URL 加载的 Document

指定 URL，可先使用 Jsoup 请求 URL，获取对应的 Document。之后，再使用 select(String cssQuery)方法定位要解析的内容。程序 5-2 为使用 Jsoup 解析 URL 加载 Document 的案例，其输出结果与程序 5-1 的输出结果一样。

程序 5-2

```
import java.io.IOException;
import org.jsoup.Jsoup;
import org.jsoup.nodes.Document;
import org.jsoup.nodes.Element;
public class JsoupParseURLDoc {
  public static void main(String[] args) throws IOException {
      //获取URL对应的Document
      Document doc = Jsoup.connect("http://www.********.com.cn/b.asp").timeout(5000).get();
      //基于CSS选择器获取元素，这里换了一种方式
      Element element = doc.select("div#w3school").get(0);
      System.out.println("输出解析的元素内容为:");
      System.out.println(element);
      //从Element提取内容(抽取一条Node对应的信息)
      String text1 = element.select("h1").text();
      //从Element提取内容(抽取一条Node对应的信息)
      String text2 = element.select("p").text();
      System.out.println("抽取的文本信息为:");
```

```
      System.out.println(text1 + "\t" + text2);
  }
}
```

3. Jsoup 遍历元素

org.jsoup.nodes.Element 类中的 select(String cssQuery)方法返回的是 Elements 对象，即匹配得到 cssQuery 对应的所有元素集合。org.jsoup.select.Elements 类继承了 ArrayList<Element>，所以，可以按照操作 ArrayList 集合的方式操作 Elements 对象，如获取某元素可使用 get(int index)方法（在程序 5-1 和程序 5-2 中皆有使用）。同样，也可以按照遍历 ArrayList 集合的方式遍历 Elements 对象。仍以某培训网站页面为例（http://www.********.com.cn/b.asp），页面内容如图 5.2 所示。如果要解析得到图 5.2 所示每个课程表的标题及每个课程表对应的 URL，则需要遍历选中的元素。

图 5.2 某培训网站页面内容

首先，利用网络抓包定位到需要解析的 HTML 片段，如程序 5-3 所示。分析程序 5-3 中的 HTML 片段，发现需要解析的每个课程表的标题及 URL 在元素<a>中，元素<a>在元素<li>中，元素<li>在 div[id=course]中。针对这种层层嵌套的结构，可以使用 Jsoup 中的遍历解析出每部分内容。具体代码如程序 5-4 所示，程序 5-4 的输出结果如图 5.3 所示。

程序 5-3

```
<div id="course">
  <ul>
    <li>
      <a href="/js/index.asp" title="JavaScript 教程">JavaScript </a></li>
    <li>
      <a href="/htmldom/index.asp" title="HTML DOM 教程">HTML DOM
```

```html
</a></li>
    <li>
      <a href="/jquery/index.asp" title="jQuery 教程">jQuery</a></li>
    <li>
      <a href="/ajax/index.asp" title="AJAX 教程">AJAX</a></li>
    <li>
      <a href="/json/index.asp" title="JSON 教程">JSON</a></li>
    <li>
      <a href="/dhtml/index.asp" title="DHTML 教程">DHTML</a></li>
    <li>
      <a href="/e4x/index.asp" title="E4X 教程">E4X</a></li>
    <li>
      <a href="/wmlscript/index.asp" title="WMLScript 教程">WMLScript
</a></li>
    </ul>
  </div>
```

程序 5-4

```java
import java.io.IOException;
import org.jsoup.Jsoup;
import org.jsoup.nodes.Document;
import org.jsoup.nodes.Element;
import org.jsoup.select.Elements;
public class JsoupParseEveryEle {
 public static void main(String[] args) throws IOException {
     //获取URL对应的Document
     Document doc = Jsoup.connect("http://www.********.com.cn/b.asp")
             .timeout(5000).get();
     //层层定位到要解析的内容,可以发现包含多个li元素
     Elements elements = doc.select("div#course").select("li");
     //遍历每一个li节点
     for (Element ele : elements) {
         //.text()为解析标签中的文本内容
         String title = ele.select("a").text();
         //.attr(String)表示获取标签内某属性的内容
         String course_url = ele.select("a").attr("href");
         System.out.println("标题为:"+title+"\tURL为:"+course_url);
     }
  }
 }
```

```
标题为:JavaScript  URL为:/js/index.asp
标题为:HTML DOM   URL为:/htmldom/index.asp
标题为:jQuery     URL为:/jquery/index.asp
标题为:AJAX       URL为:/ajax/index.asp
标题为:JSON       URL为:/json/index.asp
标题为:DHTML      URL为:/dhtml/index.asp
标题为:E4X        URL为:/e4x/index.asp
标题为:WMLScript  URL为:/wmlscript/index.asp
```

图 5.3　程序 5-4 的输出结果

4．Jsoup 获取元素的其他方法

除了 select(String cssQuery)方法，Element 类还提供了其他获取元素的方法，如表 5.7 所示。

表 5.7　获取元素的部分方法

方　　法	描　　述
Element getElementById(String id)	基于 id 获取元素
Elements getElementsByTag(String tagName)	基于标签名称获取元素集合
Elements getElementsByAttribute(String key)	基于属性名称获取元素集合
Elements getElementsByClass(String className)	基于类名获取元素集合
Elements getElementsByAttributeStarting(String keyPrefix)	通过属性前缀获取元素集合
Elements getElementsByAttributeValue(String key, String value)	基于属性和属性值获取元素集合
Elements getElementsByAttributeValueStarting(String key, String valuePrefix)	基于属性和属性值前缀获取元素集合
Elements getElementsByAttributeValueEnding(String key, String valueSuffix)	基于属性和属性值后缀获取元素集合
Elements getElementsByAttributeValueContaining(String key, String match)	基于属性与属性值（包含某字符串）获取元素集合
Elements getElementsByAttributeValueMatching(String key, String regex)	基于属性与属性值（正则表达式）获取元素集合
Elements siblingElements()	获取兄弟元素集合，如果没有返回空列表
Element nextElementSibling()	获取下一个兄弟元素，如果没有返回 null
Element previousElementSibling()	获取上一个兄弟元素，如果没有返回 null
Element firstElementSibling()	获取第一个兄弟元素
Element lastElementSibling()	获取最后一个兄弟元素
Elements children()	获取子元素集合

程序 5-5 所示为表 5.7 中一些方法的使用情况。

程序 5-5

```
        //获取 URL 对应的 Document
        Document doc = Jsoup.connect("http://www.********.com.cn/b.
asp").timeout(5000).get();
        //基于id获取元素
        Element element_id = doc.getElementById("course");
        //基于标签名称获取元素集合
        Elements element_tag = doc.getElementById("course").
getElementsByTag("a");
        //基于属性名称获取元素集合
        Elements element_A = doc.getElementById("course").
getElementsByAttribute("href");
        //通过类名获取元素集合
        Elements elements = doc.getElementsByClass("browserscripting");
        //基于属性前缀获取元素集合
        Elements element_As = doc.getElementsByAttributeStarting
("hre");
        //基于属性与属性值获取元素
        Elements element_Av = doc.getElementsByAttributeValue("id",
"tools");
        //获取兄弟元素集合
        Elements element_Se = doc.getElementById("navfirst").
siblingElements();
        //获取下一个兄弟元素
        Element element_Ns = doc.getElementById("navfirst").
nextElementSibling();
        //获取上一个兄弟元素
        Element element_Ps = doc.getElementById("navfirst").
previousElementSibling();
```

5. 支持 Xpath 语法的 JsoupXpath

Jsoup 在选择节点或元素时，支持的是 CSS 选择器，而不支持 Xpath 语法。而 JsoupXpath 则是在 Jsoup 的基础上扩展的支持 Xpath 语法的 HTML 文件解析器。

使用 JsoupXpath 解析 HTML 文件之前，需要在 Maven 工程的 pom.xml 文件中添加 JsoupXpath 对应的 dependency。

```
        <!-- https://*************.com/artifact/cn.wanghaomiao/
JsoupXpath -->
        <dependency>
            <groupId>cn.wanghaomiao</groupId>
```

```xml
        <artifactId>JsoupXpath</artifactId>
        <version>2.2</version>
    </dependency>
```

基于 pom.xml 文件的配置信息，可以下载 JsoupXpath 的 jar 包（版本为 2.2）。另外，可以看到 JsoupXpath 依赖的 jar 包有 jsoup、commons-lang、antlr4-runtime 和 slf4j-api，如图 5.4 所示。

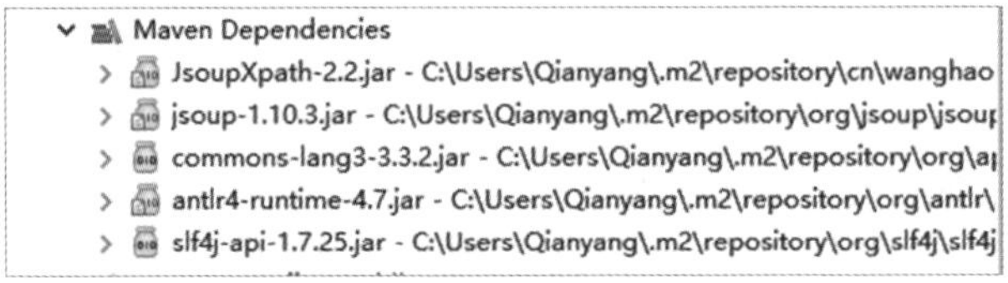

图 5.4　JsoupXpath 的 jar 包及其依赖 jar 包

在 JsoupXpath 解析 HTML 时，可以先通过 JXDocument 类中提供的 4 种方法实例化 JXDocument 对象，如程序 5-6 所示。这四种方法传递的参数分别是 Document 类型的 HTML 文档、Elements 类型的元素集合、String 类型的 HTML 字符串和 String 类型的 URL。

程序 5-6

```java
    public static JXDocument create(Document doc){
        Elements els = doc.children();
        return new JXDocument(els);
    }
    public static JXDocument create(Elements els){
        return new JXDocument(els);
    }
    public static JXDocument create(String html){
        Elements els = Jsoup.parse(html).children();
        return new JXDocument(els);
    }
    public static JXDocument createByUrl(String url){
        Elements els;
        try {
            els = Jsoup.connect(url).get().children();
        } catch (Exception e) {
            throw new XpathParserException("url ",e);
        }
        return new JXDocument(els);
    }
```

程序 5-7 展示了 JsoupXpath 解析图 5.2 中每个课程表的标题和每个课程表对应的 URL，该程序的输出结果与程序 5-4 的输出结果（见图 5.3）一致。

程序 5-7

```java
import java.util.List;
import org.seimicrawler.xpath.JXDocument;
import org.seimicrawler.xpath.JXNode;
public class JsoupXpathTest {
 public static void main(String[] args) {
     //基于URL创建JXDocument
     JXDocument jxd = JXDocument.createByUrl("http://www.********.com.cn/b.asp");
     //Xpath语句
     String str = "//*[@id='course']/ul/li/a";
     //获取节点集合
     List<JXNode> list = jxd.selN(str);
     //遍历节点
     for (int i = 0; i < list.size(); i++) {
         JXNode node = list.get(i);
         System.out.println("标题为:" + node.asElement().text() +
             "\tURL为:" + node.asElement().attr("href"));
     }
  }
 }
```

5.1.4　HtmlCleaner 解析 HTML

HtmlCleaner 是另外一款基于 Java 开发的 HTML 文档解析器，支持 XPath 语法提取 HTML 中的节点或元素。

1．jar 包的下载

在 MVNRepository 中搜索 HtmlCleaner，并使用 Eclipse 构建 Maven 工程，配置 Maven 工程中的 pom.xml 文件下载 HtmlCleaner 相关依赖 jar 包。这里以最新版的 HtmlCleaner 配置为例。

```xml
        <!-- https://*************.com/artifact/net.sourceforge.htmlcleaner/htmlcleaner -->
        <dependency>
            <groupId>net.sourceforge.htmlcleaner</groupId>
            <artifactId>htmlcleaner</artifactId>
```

```
            <version>2.22</version>
        </dependency>
        <dependency>
```

2. HtmlCleaner 类与 TagNode 类

使用 HtmlCleaner 解析 HTML 文档时，需要使用到两个类：org.htmlcleaner.HtmlCleaner 以及 org.htmlcleaner.TagNode。HtmlCleaner 类提供了实例化 HtmlCleaner 的方法和将指定类型输入（如 String 类型的 HTML 字符串、String 类型的 URL 等）转化成节点（TagNode）的方法，表 5.8 列举了 HtmlCleaner 类中的部分方法和说明。

表 5.8　HtmlCleaner 类中的部分方法和说明

方　　法	说　　明
TagNode clean(String htmlContent)	将 String 类型的 HTML 转化成 TagNode
TagNode clean(File file, String charset)	按照指定字符集将 File 中存储的 HTML 转化成 TagNode
TagNode clean(File file)	按照默认字符集将 File 中存储的 HTML 转化成 TagNode
TagNode clean(URL url, String charset)	获取 URL 对应的 HTML 内容（指定字符集），并将其转化成 TagNode
TagNode clean(URL url)	获取 URL 对应的 HTML 内容（默认字符集），并将其转化成 TagNode
TagNode clean(InputStream in, String charset)	按照指定字符集将输入流中的 HTML 内容转化成 TagNode
TagNode clean(InputStream in)	按照默认字符集将输入流中的 HTML 内容转化成 TagNode

TagNode 类提供了一系列操作节点的方法，如表 5.9 所示。

表 5.9　TagNode 类中操作节点的方法及说明

方　　法	说　　明
Object[] evaluateXPath(String xPath)	基于 Xpath 语法获取 Object 数组
String getAttributeByName(String attName)	基于属性名称获取属性值
TagNode findElementByName(String findName, boolean isRecursive)	基于标签名获取节点
CharSequence getText()	获取节点中的文本
TagNode getParent()	获取父节点
List<TagNode> getChildTagList()	获取所有子节点，返回值为集合
boolean hasChildren()	判断是否存在子节点

3. 案例讲解

下面用 HtmlCleaner 解析 W3school 页面中的数据，如图 5.5 所示。解析的内容包括课程表名称、课程表 URL、大标题和每个课程表的文本描述信息，解析过程如程序 5-8 所示。

图 5.5　HtmlCleaner 需要解析的内容

首先，使用 Jsoup 获取 String 类型 URL 对应的 HTML 内容。然后，实例化 HtmlCleaner，并基于 HtmlCleaner 类中的 clean(String htmlContent)方法将 String 类型的 HTML 转化成 TagNode。之后，采用 evaluateXPath(String xPath)方法操作 TagNode 得到 Object 数组，通过对该数组的操作便能够获得具体的字段值。

程序 5-8

```java
import java.io.IOException;
import org.htmlcleaner.HtmlCleaner;
import org.htmlcleaner.TagNode;
import org.htmlcleaner.XPatherException;
import org.jsoup.Jsoup;
import org.jsoup.nodes.Document;
public class HtmlcleanerTest1 {
  public static void main(String[] args) throws IOException,
XPatherException {
      //使用Jsoup获取HTML文件
      Document doc = Jsoup.connect("http://www.********.com.cn/b.asp").
            timeout(5000).get();
      //转化成String格式
      String html =doc.html();
      //实例化HtmlCleaner
      HtmlCleaner cleaner = new HtmlCleaner();
      //转化成TagNode
      TagNode node = cleaner.clean(html);
      //通过Xpath定位标题的位置，这里使用//h1和使用/h1的结果是一样的
       Object[] ns = node.evaluateXPath
("//div[@id='********']//h1");
      System.out.println("HTML中的标题是:\t" + ((TagNode)ns[0]).
```

```
                getText());
                    Object[]  ns1 = node.evaluateXPath("//*[@id='********']/h1");
                    System.out.println("HTML中的标题是:\t" + ((TagNode)ns1[0]).
getText());
                    //这里使用//a表示不考虑位置,如果使用/a无法获取内容
                    Object[]  ns2 = node.evaluateXPath("//*[@id='course']/ul//a");
                    for(Object on : ns2) {    //遍历获取课程名以及课程地址
                        TagNode n = (TagNode) on;
                        System.out.println("课程名为:\t" + n.getText() +
                            "\t地址为:\t" + n.getAttributeByName("href"));
                    }
                    //获取每个课程名称及其对应的简介
                    Object[]  ns3 = node.evaluateXPath("//*[@id='maincontent']
//div");
                    for (int i = 1; i < ns3.length; i++) {
                        TagNode n = (TagNode) ns3[i];
                        //获取课程名称
                        String courseName = n.findElementByName("h2", true).
getText().toString();
                        //循环遍历所有的p节点获取课程简介
                        Object[] objarrtr = n.evaluateXPath("//p");
                        String summary = "";
                        for(Object on : objarrtr) {
                            summary += ((TagNode) on).getText().toString();
                        }
                        System.out.println(courseName + "\t" + summary);
                    }
                }
            }
        }
```

图 5.6 为程序 5-8 的输出结果。

图 5.6　程序 5-8 的输出结果

5.1.5 HTMLParser 解析 HTML

HTMLParser 也是一款非常高效的 HTML 解析器，其支持 CSS 选择器提取 HTML 中的节点。HTMLParser 的版本已不再更新，但并不影响其使用。

1．jar 包的下载

在 MVNRepository 中搜索 HTMLParser，并使用 Maven 工程中的 pom.xml 文件配置 HTMLParser。

```
        <!-- https://************.com/artifact/org.htmlparser
/htmlparser -->
        <dependency>
            <groupId>org.htmlparser</groupId>
            <artifactId>htmlparser</artifactId>
            <version>2.1</version>
        </dependency>
```

2．相关方法

HTMLParser 的核心类是 org.htmlparser.Parser，其主要功能是实例化 Parser。表 5.10 列举了该类中常用的构造方法。

表 5.10　Parser 类中常用的构造方法

方　　法	说　　明
Parser()	无参数构造
Parser(Lexer lexer)	通过 Lexer 构造 Parser，在案例程序中会使用到
Parser(String resource)	给定一个 URL 或文件资源，构造 Parser
Parser(URLConnection connection)	使用 URLConnection 构造 Parser

表 5.10 中的几种构造方法能够创建 Parser 对象。在 Parser 对象创建完成后，需要执行过滤（Filter）任务，即获取包含解析内容的一系列节点，表 5.11 所示为执行过滤任务常用的过滤器。

表 5.11　常用的过滤器

方　　法	说　　明
TagNameFilter (String name)	根据标签名称进行过滤
NodeClassFilter (Class cls)	根据类名进行过滤
HasAttributeFilter (String attribute)	匹配出包含指定属性的节点
HasAttributeFilter (String attribute, String value)	匹配出包含指定属性与属性值的节点
AndFilter (NodeFilter left, NodeFilter right)	结合几种过滤条件的"与"过滤器

续表

方　法	说　明
OrFilter (NodeFilter left, NodeFilter right)	结合几种过滤条件的"或"过滤器
StringFilter (String pattern)	根据模式对象进行过滤
RegexFilter (String pattern)	根据正则表达式进行过滤
LinkStringFilter (String pattern)	判断链接中是否包含某个特定的字符串，可以用来过滤出指向某个特定网站的链接
CssSelectorNodeFilter (String selector)	根据 CSS 选择器获取节点（最常用）

完成过滤（Filter）操作之后，便要对具体的节点进行操作，常用的方法如表 5.12 所示。

表 5.12　节点操作常用方法

方　法	说　明
NodeList getChildren()	获取子节点列表
Node getParent()	获取父节点
Node getFirstChild()	获取第一个子节点
Node getLastChild()	获取最后一个子节点
Node getPreviousSibling()	获取上一个兄弟节点
Node getNextSibling()	获取下一个兄弟节点
String getText()	获取节点的文本数据（包括标签名称、属性及属性值以及标签内的数据）
String toPlainTextString()	获取节点标签内的数据，注意和 getText()的区别
String toHtml()	返回节点对应的 HTML 片段
Page getPage()	获取节点对应的 Page 对象
int getStartPosition()	获取节点在 HTML 页面中的起始位置
int getEndPosition()	获取节点在 HTML 页面中的结束位置

3．案例讲解

下面用 3 个案例程序讲解上述方法的使用情况。程序 5-9 所示为第一个案例。首先，使用 Jsoup 获取指定 URL 对应的 HTML 字符串，接着使用 Parser(Lexer lexer)构造方法创建 Parser 对象，然后结合过滤器以及操作节点的相关方法提取 HTML 中的所有链接（即 href 的属性值和链接对应的标题）。

程序 5-9

```
import java.io.IOException;
import org.htmlparser.Node;
import org.htmlparser.NodeFilter;
import org.htmlparser.Parser;
import org.htmlparser.filters.NodeClassFilter;
```

```java
import org.htmlparser.lexer.Lexer;
import org.htmlparser.tags.LinkTag;
import org.htmlparser.util.NodeList;
import org.htmlparser.util.ParserException;
import org.jsoup.Jsoup;
import org.jsoup.nodes.Document;
public class HTMLParserTest1 {
  public static void main(String[] args) throws IOException, ParserException {
        //使用Jsoup获取HTML文件
        Document doc = Jsoup.connect("http://www.********.com.cn/b.asp")
                .timeout(5000).get();
        //转化成String格式
        String html =doc.html();
        //使用lexer构造
        Lexer lexer = new Lexer(html);
        Parser parser = new Parser(lexer);
        //过滤页面中的链接标签
        NodeFilter filter = new NodeClassFilter(LinkTag.class);
        //获取匹配到的节点
        NodeList list = parser.extractAllNodesThatMatch(filter);
        //遍历每一个节点，获取链接及标题
        for(int i=0; i<list.size();i++){
            Node node = (Node)list.elementAt(i);
            System.out.println("链接为：" + ((LinkTag) node).getLink()
                + "\t标题为:" + node.toPlainTextString() );
        }
    }
}
```

图 5.7 展示了程序 5-9 的输出结果。

图 5.7　程序 5-9 的输出结果

图 5.8 所示为第二个案例所要解析的内容，对应的程序为 5-10。在程序 5-10 中，使用 Parser(String resource)构造方法实例化 Parser，并使用多个过滤器的"与"操作来获取节点链接和节点标签内的数据，其运行结果如图 5.9 所示。

图 5.8　第二个案例所要解析的内容

程序 5-10

```java
import java.io.IOException;
import org.htmlparser.Node;
import org.htmlparser.NodeFilter;
import org.htmlparser.Parser;
import org.htmlparser.filters.AndFilter;
import org.htmlparser.filters.HasAttributeFilter;
import org.htmlparser.filters.HasParentFilter;
import org.htmlparser.filters.TagNameFilter;
import org.htmlparser.tags.LinkTag;
import org.htmlparser.util.NodeList;
import org.htmlparser.util.ParserException;
public class HTMLParserTest2 {
  public static void main(String[] args) throws IOException, ParserException {
        //实例化Parser，用网页的 URL 作为参数
        Parser parser = new Parser("http://www.********.com.cn/b.asp");
        //设置网页的编码
        parser.setEncoding("gbk");
        //过滤页面中的标签
        NodeFilter filtertag= new TagNameFilter("ul");
        //父节点包含ul
        NodeFilter filterParent = new HasParentFilter(filtertag);
        //包含li标签，并且li节点中包含id属性
        NodeFilter filtername = new TagNameFilter("li");
```

```
    NodeFilter filterId= new HasAttributeFilter("id");
    //过滤器的"并"操作
    NodeFilter filter = new AndFilter(filterParent,filtername);
    NodeFilter filterfinal = new AndFilter(filter,filterId);
    //选择匹配到的内容
    NodeList list = parser.extractAllNodesThatMatch(filterfinal);
    //循环遍历
    for(int i=0; i<list.size();i++){
        //获取li的第一个子节点
        Node node = (Node)list.elementAt(i).getFirstChild();
        System.out.println( "链接为: " + ((LinkTag) node).getLink()
            +"\t标题为:" + node.toPlainTextString() );
    }
  }
}
```

```
链接为: http://www.     .com.cn/h.asp    标题为:HTML 系列教程
链接为: http://www.     .com.cn/b.asp    标题为:浏览器脚本
链接为: http://www.     .com.cn/s.asp    标题为:服务器脚本
链接为: http://www.     .com.cn/d.asp    标题为:ASP.NET 教程
链接为: http://www.     .com.cn/x.asp    标题为:XML 系列教程
链接为: http://www.     .com.cn/ws.asp   标题为:Web Services 系列教程
链接为: http://www.     .com.cn/w.asp    标题为:建站手册
```

图 5.9 程序 5-10 的输出结果

第三个案例程序为程序 5-11，解析的数据如图 5.2 所示，即每个课程表的标题和每个课程表对应的 URL。在程序 5-11 中，使用的是 Parser(URLConnection connection) 构造方法实例化 Parser。同时，基于 CSS 选择器进行过滤。使用 CSS 选择器筛选网页内容是一种非常简单且快捷的方式，这里建议读者在解析 HTML 时，也使用这种方式。程序 5-11 的输出结果与程序 5-4 的输出结果（见图 5.3）一样。

程序 5-11

```
import java.io.IOException;
import java.net.URL;
import java.net.URLConnection;
import org.htmlparser.Node;
import org.htmlparser.Parser;
import org.htmlparser.filters.CssSelectorNodeFilter;
import org.htmlparser.tags.LinkTag;
import org.htmlparser.util.NodeList;
import org.htmlparser.util.ParserException;
public class HTMLParserTest3 {
  public static void main(String[] args) throws IOException,
ParserException {
```

```java
//使用URLConnection请求数据
URL url = new URL("http://www.********.com.cn/b.asp");
URLConnection conn = url.openConnection();
Parser parser = new Parser(conn);
//使用CSS选择器进行过滤操作
CssSelectorNodeFilter divFilter = new CssSelectorNodeFilter
("#course > ul > li");
//选择匹配到的内容
NodeList list = parser.extractAllNodesThatMatch(divFilter);
//循环遍历
for(int i=0; i<list.size();i++){
    //获取li的第一个子节点
    Node node = (Node)list.elementAt(i).getFirstChild();
    System.out.println( "链接为:" + ((LinkTag) node).getLink()
            +"\t标题为:" + node.toPlainTextString() );
  }
 }
}
```

5.2 XML 解析

Jsoup 既可以解析 HTML，也可用于解析 XML。在本节中，将使用 Jsoup 解析 XML。下面以采集某网站汽车 (捷达汽车) 销量为例进行说明，请求的 URL 如下所示。

```
http://db.auto.****.com/cxdata/xml/sales/model/model1001sales.xml
```

为方便读者阅读，这里截取了部分 XML 的内容，如下所示。

```xml
<model id="1001" name="捷达">
    <sales date="2007-01-01" salesNum="14834"/>
    <sales date="2007-02-01" salesNum="9687"/>
    <sales date="2007-03-01" salesNum="18173"/>
    <sales date="2007-04-01" salesNum="18508"/>
    <sales date="2007-05-01" salesNum="19710"/>
    <sales date="2007-06-01" salesNum="20311"/>
    <sales date="2007-07-01" salesNum="17516"/>
</model>
```

如果要解析的字段为 date(销售日期)和 salesNum(销售数量)，可采用程序 5-12。由程序 5-12 可知，Jsoup 解析 XML 的方法与解析 HTML 的方法相同，皆使用 CSS 选择器选择元素。

程序 5-12

```java
import java.io.IOException;
import org.jsoup.Jsoup;
import org.jsoup.nodes.Document;
import org.jsoup.nodes.Element;
import org.jsoup.select.Elements;
public class JsoupXML {
 public static void main(String[] args) throws IOException {
     //获取URL对应的HTML内容
     String url = "http://db.auto.****.com/cxdata/xml/sales/model/model1001sales.xml";
     Document doc = Jsoup.connect(url).timeout(5000).get();
     //Jsoup选择器解析
     Elements sales_ele = doc.select("sales");
     for (Element elem:sales_ele) {
         int salesnum=Integer.valueOf(elem.attr("salesnum"));
         String date = elem.attr("date");
         System.out.println("月份:" + date + "\t销量:" + salesnum);
     }
  }
 }
```

图 5.10 为程序 5-12 的输出结果。

```
月份:2007-01-01    销量:14834
月份:2007-02-01    销量:9687
月份:2007-03-01    销量:18173
月份:2007-04-01    销量:18508
月份:2007-05-01    销量:19710
月份:2007-06-01    销量:20311
月份:2007-07-01    销量:17516
月份:2007-08-01    销量:17535
月份:2007-09-01    销量:17743
月份:2007-10-01    销量:15255
月份:2007-11-01    销量:17250
月份:2007-12-01    销量:14609
月份:2008-01-01    销量:25126
月份:2008-02-01    销量:9077
月份:2008-03-01    销量:20396
```

图 5.10　程序 5-12 的输出结果

5.3　JSON 解析

5.3.1　JSON 校正

使用网络爬虫向服务器发送请求时，服务器经常返回的数据是包含 JSON 的字符串，如下所示。

```
jQuery16({
    "id":"07",
    "language": "C++",
    "edition": "second",
    "author": "E.Balagurusamy"
})
```

上述字符串虽包含 JSON，但并不能直接使用 org.json、Gson 和 Fastjson 等工具进行直接解析，因其在头部和尾部包含多余的字符（"jQuery16("与"）"）。为使上述字符串可以直接解析，需要对其进行预处理（掐头去尾）操作，将其转化为标准的 JSON 字符串。程序 5-13 给出了一种处理方式，其输出结果如图 5.11 所示。

程序 5-13

```
//拼接 JSON 字符串
String json = "jQuery16({\"id\":\"07\","
        + "\"language\": \"C++\",\"edition\": \"second\","
        + "\"author\": \"E.Balagurusamy\"})";
//掐头去尾操作
String arr = json.split("\\(")[1];
System.out.println(arr.substring(0,arr.length() - 1));
```

```
{"id":"07","language": "C++","edition": "second","author":
"E.Balagurusamy"}
```

图 5.11 程序 5-13 的输出结果

针对已经处理好的字符串，可以复制到 JSON 在线校准网站（注：该网站在分析 JSON 时经常用到）。

JSON 在线校准网站是一款集格式化和验证功能于一体的工具，使用该工具可以实现 JSON 数据的在线验证、编辑和格式化。图 5.12 所示为图 5.11 所示 JSON 字符串的在线校准结果。

图 5.12 图 5.11 所示 JSON 字符串在线校准结果

5.3.2 org.json 解析 JSON

1. jar 的下载

org.json 是 Java 中常用的一款 JSON 解析工具。使用前，在 MVNRepository 中搜索 org.json，并使用 Maven 工程中的 pom.xml 文件配置 org.json。

```xml
<!-- https://************.com/artifact/org.json/json -->
<dependency>
    <groupId>org.json</groupId>
    <artifactId>json</artifactId>
    <version>20180130</version>
</dependency>
```

org.json 解析 JSON，常用的两个类分别是 JSONObject 和 JSONArray，下面将详细介绍这两个类的使用情况。

2. JSONObject 类

JSONObject 类的功能是处理 JSON 对象，该类中包括一系列实例化 JSONObject 对象的构造方法。表 5.13 所示为 JSONObject 类中的部分构造方法及说明。

表 5.13　JSONObject 类中的一些构造方法

方　　法	说　　明
JSONObject()	实例化空的 JSONObject 对象
JSONObject(String source)	基于 JSON 字符串实例化 JSONObject 对象
JSONObject(JSONTokener x)	基于 JSONTokener 对象实例化 JSONObject 对象
JSONObject(Map<?, ?> m)	基于 Map 集合实例化 JSONObject 对象
JSONObject(Object bean)	基于 Bean 对象实例化 JSONObject 对象
JSONObject(Object object, String names[])	基于反射机制（公有变量）实例化 JSONObject 对象

同时，JSONObject 类中还提供了添加数据和获取数据的方法等，如表 5.14 所示。

表 5.14　JSONObject 类中添加数据和获取数据的部分方法

方　　法	说　　明
JSONObject accumulate(String key, Object value)	JSONObject 对象添加数据
JSONObject append(String key, Object value)	JSONObject 对象添加数据
Object get(String key)	获取 JSONObject 对象中键 key 对应的 value，返回的类型为 Object
boolean getBoolean(String key)	判断 JSONObject 对象中键 key 是否存在

续表

方　　法	说　　明
String getString(String key)	获取 JSONObject 对象中键 key 对应的 value，返回的类型为 String

程序 5-14 所示为表 5.13 和表 5.14 中部分方法的使用情况。其中，BookModel 和 BookModel2 为创建的 JavaBean，两者唯一的区别是 BookModel 中的变量全部使用 private 修饰，BookModel2 中的变量全部使用 public 修饰。图 5.13 所示为程序 5-14 的输出内容。

程序 5-14

```java
import java.util.HashMap;
import java.util.Map;
import org.json.JSONObject;
import org.json.JSONTokener;
import com.model.BookModel;
import com.model.BookModel2;
public class OrgJsonBeanToObject {
 public static void main(String[] args) {
     JSONObject beanToJson= beanToJson();
     System.out.println(beanToJson);
     JSONObject mapToJson = mapToJson();
     System.out.println(mapToJson);
     JSONObject jSONTokenerToJson = jSONTokenerToJson();
     System.out.println(jSONTokenerToJson);
     JSONObject namesToJson = namesToJson();
     System.out.println(namesToJson);
     JSONObject accumulateToJson = accumulateToJson();
     System.out.println(accumulateToJson);
     JSONObject appendToJson = appendToJson();
     System.out.println(appendToJson);
 }
 /***
  * 使用JSONObject(Object bean)构造方法将Bean对象转化成JSONObject
  */
 public static JSONObject beanToJson (){
     BookModel book = new BookModel();
     book.setId("07");
     book.setLanguage("Java");
     book.setEdition("third");
     book.setAuthor("Herbert Schildt");
```

```java
        //使用JSONObject(Object bean)构造方法
        return new JSONObject(book);
    }
    /***
     * 使用JSONObject(Map<?, ?> m)构造方法将Map集合数据转化成JSONObject输出
     */
    public static JSONObject mapToJson (){
        Map<String,String> bookmap = new HashMap<String, String>();
        bookmap.put("id", "07");
        bookmap.put("author", "Herbert Schildt");
        bookmap.put("edition", "third");
        bookmap.put("language", "Java");
        //使用JSONObject(Object bean)构造方法
        return new JSONObject(bookmap);
    }
    /***
     * 使用JSONObject(JSONTokener x)构造方法将Map集合数据转化成JSONObject
输出
     */
    public static JSONObject jSONTokenerToJson (){
        /***
         * JSONTokener类中的构造方法有:
         * 1. JSONTokener(Reader reader)
         * 2. JSONTokener(InputStream inputStream)
         * 3. JSONTokener(String s)(本案例中使用的是这种)
         */
        String jsonStr = "{\"id\":\"07\",\"language\": \"C++\",\
"edition\": \"second\",\"author\": \"E.Balagurusamy\"}";
        JSONTokener jsonTokener = new JSONTokener(jsonStr);
        return new JSONObject(jsonTokener);
    }
    /***
     * 使用JSONObject(Object object, String names[])构造方法转化成
JSONObject输出
     */
    public static JSONObject namesToJson (){
        BookModel2 book = new BookModel2();
        book.setId("07");
        book.setLanguage("Java");
        book.setEdition("third");
        book.setAuthor("Herbert Schildt");
        String names[] = {"id","language","edition","author"};
        return new JSONObject(book,names);
```

```java
    }
    /***
     * 使用JSONObject(String source)构造方法转化成JSONObject输出
     */
    public static JSONObject stringToJson (){
        //JSON字符串
        String json = "{\"id\":\"07\",\"language\": \"C++\",\"edition\": \"second\",\"author\": \"E.Balagurusamy\"}";
        return new JSONObject(json);
    }
    /***
     * JSONObject accumulate(String key, Object value)方法的使用
     */
    public static JSONObject accumulateToJson (){
        JSONObject jsonObject = new JSONObject();
        jsonObject.accumulate("id", "07");
        jsonObject.accumulate("language", "C++");
        jsonObject.accumulate("edition", "second");
        jsonObject.accumulate("author", "E.Balagurusamy");
        return jsonObject;
    }
    /***
     * JSONObject append(String key, Object value)方法的使用
     */
    public static JSONObject appendToJson (){
        JSONObject jsonObject = new JSONObject();
        jsonObject.append("id", "07");
        jsonObject.append("language", "C++");
        jsonObject.append("edition", "second");
        jsonObject.append("author", "E.Balagurusamy");
        return jsonObject;
    }
}
```

```
{"author":"Herbert Schildt","edition":"third","language":"Java","id":"07"}
{"edition":"third","language":"Java","id":"07","author":"Herbert Schildt"}
{"author":"E.Balagurusamy","edition":"second","language":"C++","id":"07"}
{"edition":"third","language":"Java","id":"07","author":"Herbert Schildt"}
{"author":"E.Balagurusamy","edition":"second","language":"C++","id":"07"}
{"author":["E.Balagurusamy"],"edition":["second"],"language":["C++"],"id":["07"]}
```

图 5.14 程序 5-13 的输出结果

在解析数据时，常用 getString(String key)方法获取 JSON 数据中 key 值对应的 value 值，如程序 5-15 所示。图 5.14 所示为程序 5-15 的输出结果。

程序 5-15

```java
import org.json.JSONArray;
import org.json.JSONObject;
public class OrgJsonObjectParseTest {
  public static void main(String[] args) {
      //json对象
      String json = "{\"id\":\"07\",\"language\": \"C++\",\"edition\": \"second\",\"author\": \"E.Balagurusamy\"}";
      //JSONObject构造对象
      JSONObject jsonObject = new JSONObject(json);
      //获取所有的key值
      JSONArray keys = jsonObject.names();
      System.out.println(keys);
      String id=jsonObject.getString("id");   //获取key对应的value
      String language=jsonObject.getString("language");
      String edition=jsonObject.getString("edition");
      //输出结果
      System.out.println(id + "\t" + language + "\t" + edition);
  }
}
```

```
01      Java     third
07      C++      second
```

图 5.14　程序 5-15 的输出结果

3. JSONArray 类

JSONArray 类的功能是解析 JSON 数组，该类中包括一些实例化 JSONArray 对象的构造方法、获取指定 JSONObject 对象的方法等。程序 5-16 所示为 JSONArray 解析 JSON 数组的案例。

程序 5-16

```java
import org.json.JSONArray;
import org.json.JSONObject;
public class OrgJsonArrayParseTest {
  public static void main(String[] args) {
      //JSON数组
      String json = "[{\"id\":\"01\",\"language\": \"Java\",\"edition\":\"third\",\"author\": \"Herbert Schildt\"},{\"id\":\"07\", \"language\": \"C++\",\"edition\": \"second\",\"author\": \"E.Balagurusamy\"}]";
      /*
```

```
     * 转化成JSONArray对象
     * 使用的是JSONArray(String source)构造方法
     **/
    JSONArray jsonarray = new JSONArray(json);
    for (int i = 0; i < jsonarray.length(); i++) {
        /*
         * 获取指定JSON对象
         * 使用的是JSONObject getJSONObject(int index)方法
         **/
        JSONObject jsonobj = jsonarray.getJSONObject(i);
        String id = jsonobj.getString("id");
        String language = jsonobj.getString("language");
        String edition = jsonobj.getString("edition");
        //输出解析的结果
        System.out.println(id + "\t" + language + "\t" + edition);
    }
  }
}
```

图 5-15 为程序 5.16 的输出结果。

```
01      Java    third
07      C++     second
```

图 5.15 程序 5-16 的输出结果

5.3.3 Gson 解析 JSON

1. jar 包的下载

Gson 是 Google 提供的处理 JSON 数据的 Java 类库，主要用于转换 Java 对象和 JSON 对象。在 Maven 工程的 pom.xml 文件中配置最新的 Gson。

```
        <!-- https://*************.com/artifact/com.google.code.gson/gson -->
        <dependency>
            <groupId>com.google.code.gson</groupId>
            <artifactId>gson</artifactId>
            <version>2.8.5</version>
        </dependency>
```

2. 解析 JSON 对象

在网络爬虫中，主要利用 Gson 将复杂的 JSON 数据转化成 Java 对象。程序 5-17

给出了 5.3.2 节中用到 BookModel 类（JavaBean），该类封装了需要解析的字段，如书名 id 和版本 edition 等，可以基于该类实例化 Java 对象。

程序 5-17

```java
public class BookModel {
 private String id;
 private String language;
 private String edition;
 private String author;
 public String getId() {
     return id;
 }
 public void setId(String id) {
     this.id = id;
 }
 public String getLanguage() {
     return language;
 }
 public void setLanguage(String language) {
     this.language = language;
 }
 public String getEdition() {
     return edition;
 }
 public void setEdition(String edition) {
     this.edition = edition;
 }
 public String getAuthor() {
     return author;
 }
 public void setAuthor(String author) {
     this.author = author;
 }
}
```

在使用 Gson 解析数据时，需要对其实例化，具体操作方式有两种。

```java
Gson gson = new Gson();                        //实例化操作方式 1
Gson gson = new GsonBuilder().create();        //实例化操作方式 2
```

在实例化 Gson 之后，可以使用 Gson 类中提供的方法将 JSON 字符串转化成指定的 Java 对象，具体方法如下所示。

```
<T> T fromJson(String json, Class<T> classOfT)
```

程序 5-18 所示为使用 Gson 解析 JSON 对象的一个案例,该程序的输出结果如图 5.16 所示。

程序 5-18

```java
import com.google.gson.Gson;
import com.model.BookModel;
public class GsonParseObjectTest {
  public static void main(String[] args) {
     //JSON对象
     String json = "{\"id\":\"07\",\"language\": \"C++\",\"edition\": \"second\",\"author\": \"E.Balagurusamy\"}";
     Gson gson = new Gson();   //实例化操作
     //转化成Java对象
     BookModel model = gson.fromJson(json, BookModel.class);
     //输出数据
     System.out.println(model.getId() + "\t" + model.getLanguage() + "\t" + model.getEdition());
  }
}
```

```
07        C++       second
```

图 5.16 程序 5-18 的输出结果

3. 解析 JSON 数组

使用 Gson 解析 JSON 数组时,需要使用 Gson 类中提供的方法将 JSON 数组转化成指定类型的集合,方法如下:

```
<T> T fromJson(String json, Type typeOfT)
```

该方法的输入参数类型为 **String** 和 **Type**。其中,引入 Type 类型是为了支持对泛型的操作。在 Gson 包中,可以使用 TypeToken 类来创建 Type 对象。程序 5-19 给出了 Gson 解析 JSON 数组的操作,其输出结果与程序 5-16 的输出结果一样(见图 5.15)。

程序 5-19

```java
import java.lang.reflect.Type;
import java.util.List;
import com.google.gson.Gson;
import com.google.gson.reflect.TypeToken;
```

```java
import com.model.BookModel;
public class GsonParseArrayTest {
    public static void main(String[] args) {
        //JSON数组
        String json = "[{\"id\":\"01\",\"language\": \"Java\",\"edition\": \"third\",\"author\": \"Herbert Schildt\"},{\"id\":\"07\", \"language\": \"C++\",\"edition\": \"second\",\"author\":\"E.Balagurusamy\"}]";
        Gson gson = new Gson();   //实例化Gson
        //TypeToken操作,可支持类型包括泛型
        Type listType = new TypeToken <List<BookModel>>(){}.getType();
        //转化成集合
        List<BookModel> listmodel = gson.fromJson(json, listType);
        //输出数据
        for (BookModel model : listmodel) {
            System.out.println(model.getId() + "\t" + model.getLanguage() + "\t" + model.getEdition());
        }
    }
}
```

4. 解析复杂嵌套式的 JSON 数据

在网络爬虫中,经常遇到一些复杂嵌套式的 JSON 数据,如程序 5-20 所示的 JSON 数据。

程序 5-20

```
{
    "goodRateShow":99,
    "poorRateShow":1,
    "poorCountStr":"500+",
    "book": [
        {
            "id":"01",
            "language": "Java",
            "edition": "third",
            "author": "Herbert Schildt"
        },
        {
            "id":"07",
            "language": "C++",
            "edition": "second",
```

```
        "author": "E.Balagurusamy"
    }]
}
```

解析这种复杂嵌套式的 JSON 数据，只需按照 JSON 的层次结构创建相应的 JavaBean。如创建一个 BookSummaryModel 类，如程序 5-21 所示。

程序 5-21

```java
import java.util.List;

public class BookSummaryModel {
    private String goodRateShow;
    private String poorRateShow;
    private String poorCountStr;
    private List<BookModel> book;
    public String getGoodRateShow() {
        return goodRateShow;
    }
    public String getPoorRateShow() {
        return poorRateShow;
    }
    public String getPoorCountStr() {
        return poorCountStr;
    }
    public List<BookModel> getBook() {
        return book;
    }
}
```

基于创建的 BookSummaryModel 类和 BookModel 类，便可以使用 Gson 解析程序 5-20 所示的 JSON 数据，如程序 5-22 所示。图 5.17 所示为程序 5-22 的输出结果。

程序 5-22

```java
import java.util.List;
import com.google.gson.Gson;
import com.model.BookModel;
import com.model.BookSummaryModel;
public class GsonParseComplexTest {
    public static void main(String[] args) {
        //复杂嵌套式的JSON数据
```

```
        String json = "{\"goodRateShow\":99,\"poorRateShow\":1,
\"poorCountStr\":\"500+\",\"book\": [{\"id\":\"01\",\"language\":
 \"Java\",\"edition\": \"third\",\"author\": \"Herbert Schildt\"},
{\"id\":\"07\", \"language\": \"C++\",\"edition\": \"second\",
\"author\": \"E.Balagurusamy\"}]}";
        Gson gson = new Gson();     //初始化操作
        BookSummaryModel smodel = gson.fromJson(json,
BookSummaryModel.class);          //转化成Java对象
        //从对象中取得集合
        List<BookModel> listmodel = smodel.getBook();
        //输出数据
        for (BookModel model : listmodel) {
            System.out.println(smodel .getGoodRateShow() +
                 "\t"+ smodel .getPoorCountStr() + "\t" + model.getId() +
                 "\t" + model.getLanguage() + "\t" + model.getEdition());
        }
    }
 }
```

| 99 | 500+ | 01 | Java | third |
| 99 | 500+ | 07 | C++ | second |

图 5.17　程序 5-22 的输出结果

另外，Gson 还提供了其他 JSON 处理的方法，如 toJson()。有兴趣的读者，可以通过访问 Google 提供的官方文档进行学习。

5.3.4　Fastjson 解析 JSON

1．jar 包的下载

Fastjson 是阿里巴巴公司基于 Java 语言开发的高性能且功能完善的 JSON 操作类库。基于 Maven 工程的 pom.xml 文件下载最新的 Fastjson 包。

```
<!-https://*************.com/artifact/com.alibaba/fastjson →
<dependency>
    <groupId>com.alibaba</groupId>
    <artifactId>fastjson</artifactId>
    <version>1.2.47</version>
</dependency>
```

2. 解析 JSON 对象和 JSON 数组

Fastjson 解析 JSON 数据的方式与 Gson 类似，即将 JSON 数据转化成 JavaBean 对象。下面提供了解析 JSON 对象和 JSON 数组的常用方法。

```
//操作 JSON 对象的方法
<T> T parseObject(String text, Class<T> clazz)
//操作JSON数组的方法
<T> List<T> parseArray(String text, Class<T> clazz)
//基于泛型的方式，可操作JSON数组
<T> T parseObject(String text, TypeReference<T> type, Feature…features)
```

程序 5-23 所示为 Fastjson 解析 JSON 对象的一个案例。

程序 5-23

```
import com.alibaba.fastjson.JSON;
import com.model.BookModel;
public class FastjsonParseIObjectTest {
  public static void main(String[] args) {
    //JSON对象
    String json = "{\"id\":\"07\",\"language\": \"C++\",\"edition\":\"second\",\"author\": \"E.Balagurusamy\"}";
    //使用Fastjson解析JSON对象
    BookModel model = JSON.parseObject(json, BookModel.class);
    //输出解析结果
    System.out.println(model.getId() + "\t" + model.getLanguage() +
            "\t" + model.getEdition());
  }
}
```

程序 5-24 所示为 Fastjson 解析 JSON 数组的一个案例。

程序 5-24

```
import java.util.List;
import com.alibaba.fastjson.JSON;
import com.alibaba.fastjson.TypeReference;
import com.model.BookModel;
public class FastjsonParseIArrayTest {
  public static void main(String[] args) {
    //JSON数组
```

```
        String json = "[{\"id\":\"01\",\"language\": \"Java\",
\"edition\": \"third\",\"author\": \"Herbert Schildt\"},{\"id\":
\"07\", \"language\": \"C++\",\"edition\": \"second\",\"author\":
\"E.Balagurusamy\"}]";
        //使用Fastjson解析JSON数组
        List<BookModel> listmodel = JSON.parseObject(json, new
TypeReference<List<BookModel>>(){}); //第一种方式
        //第二种方式
        //List<BookModel> listmodel = JSON.parseArray(json, BookModel.
class);
        //输出数据
        for (BookModel model : listmodel) {
            System.out.println(model.getId() + "\t" + model.
getLanguage() +
                    "\t" + model.getEdition());
        }
    }
}
```

由程序 5-23 和程序 5-24 可以看出 Fastjson 是一款非常简单且实用的 JSON 解析工具。

5.3.5 网络爬虫实战演练

为提升实战能力,下面以一个实际网站为例,介绍 JSON 数据的解析。

图 5.18 所示为本案例所用网址页面部分内容,本案例的目标是采集该页面中的评论信息,即评论的 id、评论的菜品、评论的详细内容、评论的时间和评论作者的名称。确定需要采集的字段后,可按照以下步骤编写程序。

图 5.18　本案例所用网址页面部分内容

第一步：构建 CommentModel 类（JavaBean），如程序 5-25 所示。

程序 5-25

```java
public class CommentModel {
    private String CommentId;      //评论的id
    private String ItemId;         //评论的菜品
    private String Content;        //评论的内容
    private String CreateTime;     //评论的时间
    private String OpenUserName;//评论作者的名称
    public String getCommentId() {
        return CommentId;
    }
    public void setCommentId(String commentId) {
        CommentId = commentId;
    }
    public String getItemId() {
        return ItemId;
    }
    public void setItemId(String itemId) {
        ItemId = itemId;
    }
    public String getContent() {
        return Content;
    }
    public void setContent(String content) {
        Content = content;
    }
    public String getCreateTime() {
        return CreateTime;
    }
    public void setCreateTime(String createTime) {
        CreateTime = createTime;
    }
    public String getOpenUserName() {
        return OpenUserName;
    }
    public void setOpenUserName(String openUserName) {
        OpenUserName = openUserName;
    }
}
```

第二步：基于网络抓包分析该网页中评论数据对应的真实请求 URL。图 5.19 所示为谷歌浏览器抓包取得的数据。

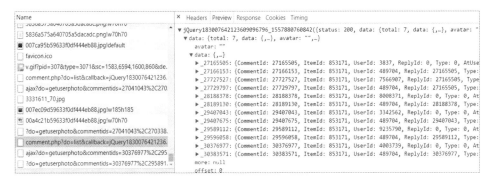

图 5.19　谷歌浏览器抓包取得的数据

如下为网络抓包获取的 URL。

```
http://www.******.com/comment.php?do=list&callback=jQuery18307729
686762719652_1543238335167&channel=recipe&item=853171&sort=desc&page
=1&size=5&comment_id=0&cate=0&purify=common&_=1543238335557
```

第三步：对网络抓包返回的字符串，执行掐头去尾操作，将其处理成标准的 JSON 字符串。这里需要使用到在线 JSON 校准网站。图 5.20 所示为 JSON 字符串校准结果。

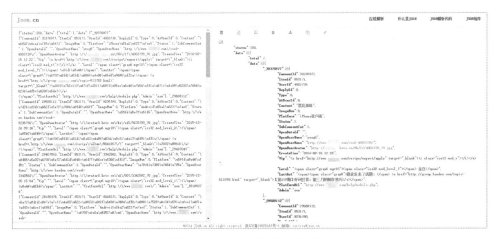

图 5.20　JSON 字符串校准结果

第四步：使用 Httpclient 或 URLConnection 请求评论对应的真实 URL，获取服务器返回的字符串。针对已获取的字符串，在程序中进行掐头去尾操作，使其转化成易于解析的标准 JSON 字符串，这里经常使用到正则表达式。最后，利用 JSON 解析工具（如 Fastjson）进行解析。

程序 5-26 所示为解析该网站评论内容的完整操作。

程序 5-26

```java
import java.io.IOException;
import java.util.List;
import org.apache.http.HttpEntity;
import org.apache.http.HttpResponse;
import org.apache.http.client.ClientProtocolException;
import org.apache.http.client.HttpClient;
import org.apache.http.client.methods.HttpGet;
import org.apache.http.impl.client.HttpClients;
import org.apache.http.util.EntityUtils;
import com.alibaba.fastjson.JSON;
import com.model.CommentModel;
public class Demo {
  private static HttpClient httpClient = HttpClients.custom().build(); //初始化HttpClient
    public static void main(String[] args) throws Exception {
        String url = " http://www.******.com/comment.php?do=list&callback=jQuery18307729686762719652_1543238335167&channel=recipe&item=853171&sort=desc&page=1&size=5&comment_id=0&cate=0&purify=common&_=1543238335557";
        //获取JSON数据
        String jsonstring = getJson(url);
        //解析JSON数据
        List<CommentModel> datalist = parseData(jsonstring);
        //输出数据
        for (CommentModel comm : datalist) {
            System.out.println(comm.getCommentId() + "\t" + comm.getItemId() + "\t" + comm.getContent());
        }
    }
    //获取JSON内容
    public static String getJson(String url) throws ClientProtocolException, IOException{
        HttpGet httpget = new HttpGet(url); //使用的请求方法
        //发出GET请求
```

```java
        HttpResponse response = httpClient.execute(httpget);
        //获取网页内容流
        HttpEntity httpEntity = response.getEntity();
        //以字符串的形式(需设置编码)
        String entity = EntityUtils.toString(httpEntity, "gbk");
        //关闭内容流
        EntityUtils.consume(httpEntity);
        return entity;              //返回JSON
    }
    //解析JSON内容
    public static List<CommentModel> parseData (String json) throws Exception{
        json = decode(json);    //将unicode码转化为中文
        //使用分割及正则取代,处理成标准JSON数组
        String jsondata = "{"+json.split("data\":\\{")[2].split("\"avatar")[0].replaceAll("\"_\\d*[0-9]\":", "");
        String jsonStr = jsondata.substring(0, jsondata.length()-2);
        //将JSON数组解析成对象集合
        List<CommentModel> datalis = JSON.parseArray("["+jsonStr.substring(1,jsonStr.length())+"]", CommentModel.class);
        return datalis;
    }
    //将unicode码转化为中文
    public static String decode(String unicodeStr) {
        if (unicodeStr == null) {
            return null;
        }
        StringBuffer retBuf = new StringBuffer();
        int maxLoop = unicodeStr.length();
        for (int i = 0; i < maxLoop; i++) {
            if (unicodeStr.charAt(i) == '\\') {
                if ((i < maxLoop - 5)
                        && ((unicodeStr.charAt(i + 1) == 'u') || (unicodeStr
                                .charAt(i + 1) == 'U')))
                    try {
                        retBuf.append((char) Integer.parseInt(
                                unicodeStr.substring(i+2,i+6),16));
                        i += 5;
```

```
                } catch (NumberFormatException
localNumberFormatException) {
                    retBuf.append(unicodeStr.charAt(i));
                }
            else
                retBuf.append(unicodeStr.charAt(i));
        } else {
            retBuf.append(unicodeStr.charAt(i));
        }
    }
    return retBuf.toString();
  }
}
```

图 5.21 所示为程序 5-26 解析得到的评论内容。

```
30376977    853171   潭亮美味
29589112    853171   紫菜是干的还是
29407043    853171   超市有干贝和海蜇卖?
28188378    853171   干贝虾米一般都是咸的,要用水多泡会,泡软
27165505    853171   食材丰富--口感也丰富!
30383571    853171   @<a href="http://www.█████.com/cook-4003739/" target="_blank">yxeg5</a> 感谢你的分享。
29596058    853171   @<a href="http://www.█████.com/cook-9235790/" target="_blank">嗡平凶</a> 是干的,要冲洗一下。
29407675    853171   @<a href="http://www.█████.com/cook-3342562/" target="_blank">秋玉的美</a> 商店里有同上也有。
28189130    853171   @<a href="http://www.█████.com/cook-8008371/" target="_blank">月上荒城6</a> 我买的这种不是那种很硬的,很多盐的
27729797    853171   @<a href="http://www.█████.com/cook-7566907/" target="_blank">haodou8704818142</a> 我在厦门,漳州咗的,
27727527    853171   @<a href="http://www.█████.com/cook-489704/" target="_blank">晓红</a> 和我们的配料不一样
27166153    853171   @<a href="http://www.█████.com/cook-3837/" target="_blank">爱跳舞的老太</a> 姐是这儿的人,不知我这样象不对吗?
```

图 5.21　程序 5-26 的解析得到的评论内容

值得注意的是 Httpclient 请求 URL 得到的字符串中,中文采用的编码为 Unicode 码。故在解析之前需将其转化成中文字符。另外,一般情况下,只知道一个菜谱的 id (如 853171),此时该如何拼接请求的 URL 呢?基于网络抓包发现真实 URL 中包含的 callback=jQuery18307729686762719652_1543238335167 以及&_=1543238335557 参数。实际上,两个参数是动态变化的,但将这两个参数从 URL 中删除,依旧可以获取 JSON 数据。删除参数后的 URL 为 http://www.******.com/comment.php?do=list&channel=recipe&item=853171…。

因此,如果给定另外一个菜谱,id 为 344953,便可有规律地拼接其评论内容对应的 URL。id 为 344953 对应的 URL 为 http://www.******.com/comment.php?do=list&channel=recipe&item=344953…。

针对评论存在多页的情况,可以调整 URL 中的 page 参数获取多页数据。如 id 为 344953 的菜谱对应的第二页评论的 URL 地址为 http://www.******.com/comment.php?do=list&channel=recipe&item=344953&sort=desc&page=1…。

5.4　本章小结

HTML、XML 以及 JSON 是网络爬虫中经常遇到的几种数据格式。针对每种格式的数据，本章都介绍了具体的解析工具。读者在阅读完本章之后，对这些工具的使用情况能够有比较深入的了解。同时，读者也可以寻找一些自己感兴趣的网站，利用本章介绍的工具，进行网站内容的解析。

第 6 章

网络爬虫数据存储

6.1 输入流与输出流

在网络爬虫中，经常需要读取数据和写入数据。例如，将存储在 txt 文档中的待采集 URL 列表读入程序；将采集的图片、PDF 和压缩包等文件写入指定目录。这些操作，都依赖于输入流与输出流。在本节中，将主要介绍 Java 中输入流与输出流的使用情况。

6.1.1 简介

输入流和输出流中的"流"的概念可以理解为从源到目的地的字节序列。数据的读写是通过字节序列的"流动"实现的。在 Java 中，输入流是能够读入一个字节序列的对象，而输出流是能够写入一个字节序列的对象。图 6.1 所示为输入流和输出流的作用。

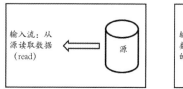
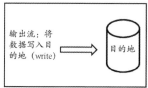

图 6.1 输入流和输出流的作用

Java 的 java.io 包提供了大量的流类，其中包括获取文件本身信息的 File 类、面向字节的输入类（InputStream、FileInputStream 等）与输出类（OutputStream、FileOutputStream 等）、面向字符的输入类（Reader、FileReader 等）与输出类（Writer、FileWriter 等）。

6.1.2 File 类

File 类主要用来获取文件自身的一些信息，包括文件所在的目录、文件是否可读、文件是否存在、文件的长度等，不会涉及文件的具体读写操作，但创建的 File 对象，可以作为字节流或字符流的输入。

1. File 类中的方法

表 6.1 所示为创建 File 对象的方法。

表 6.1 创建 File 对象的方法

方　　法	说　　明
File(String pathname)	基于一个路径创建 File 对象
File(String parent, String child)	基于目录和子文件创建 File 对象
File(File parent, String child)	基于父 File 对象和子文件创建新的 File 对象

程序 6-1 所示为表 6.1 所示方法使用情况的案例。

程序 6-1

```java
import java.io.File;
import java.net.URISyntaxException;
public class FileCreateTest {
 public static void main(String[] args) throws URISyntaxException {
    //创建File对象
    File file1 = new File("data","1.txt");
    System.out.println(file1.exists());  //判断文件是否存在
    //创建File对象
    File file2 = new File(new File("data"),"1.txt");
    System.out.println(file2.exists());
    //创建File对象
    File file3 = new File("data/1.txt");
    System.out.println(file3.exists());
 }
}
```

2. File 类基本操作方法

创建 File 对象后，可以使用 File 类中的一些方法获取文件的信息。由于 File 类中的基本操作方法较多，这里只列举部分操作方法的使用，如表 6.2 所示。

表 6.2 File 类中的部分基本操作方法

方　　法	说　　明
String getName()	获取文件名称
String getParent()	获取文件的父目录
File getParentFile()	获取文件的父目录，返回为 File 对象
String getPath()	获取文件路径名称
boolean isAbsolute()	是否为绝对路径

续表

方　　法	说　　明
String getAbsolutePath()	获取绝对路径
boolean canRead()	判断文件是否可读
boolean exists()	判断文件是否存在
boolean isDirectory()	判断路径是否为一个目录
boolean isFile()	判断文件是否为一个普通文件，而不是目录
long length()	获取文件长度
String[] list()	用字符串形式返回目录下的所有文件
File[] listFiles()	用 File 对象返回目录下的所有文件
String[] list(FilenameFilter filter)	用字符串形式返回目录下指定类型的文件

在程序 6-2 中，演示了表 6.2 中的一些操作方法的使用情况。

程序 6-2

```java
import java.io.File;
public class FileTest {
 public static void main(String[] args) {
     File root = new File("data/");
     //判断文件是否为一个目录
     Boolean is_directory = root.isDirectory();
     System.out.println("文件是否为目录:" + is_directory);
     //判断文件是否存在
     System.out.println("文件是否存在:" + root.exists());
     //判断文件是否为一个普通文件
     System.out.println("文件是否为普通文件:" + root.isFile());
     //判断是否为绝对路径
     System.out.println("文件是否为绝对路径:" + root.isAbsolute());
     //如果是一个目录
     if (is_directory) {
         //获取目录下的所有文件和目录的绝对路径，得到的是File数组
         File[] files = root.listFiles();
         //循环数组
         for ( File file : files ){
             System.out.println("============");
             System.out.println("文件名称为:" + file.getName());
             System.out.println("文件可读否:" + file.canRead());
             System.out.println("文件的父目录为:" + file.getParent());
             System.out.println("绝对路径:" + file.getAbsolutePath());
             System.out.println("相对路径:" + file.getPath());
```

```
            System.out.println("文件的长度为:" + file.length());
            System.out.println("============");
        }
    }
  }
}
```

程序 6-2 的输出结果如图 6.2 所示。从输出结果中可以发现，"data/" 并非绝对路径，并且这个文件是一个目录而不是一个普通文件。在该目录下存在名称为 1.txt 和 2.txt 的两个子文件，而且程序获得了这两个子文件的相对路径、绝对路径及文件的长度。

```
文件是否为目录:true
文件是否存在:true
文件是否为普通文件:false
文件是否为绝对路径:false
============
文件名称为:1.txt
文件可读否:true
文件的父目录为:data
绝对路径:F:\program_work\java_work\CSDNCourse\FileProcessInCrawler\data\1.txt
相对路径:data\1.txt
文件的长度为:37
============
============
文件名称为:2.txt
文件可读否:true
文件的父目录为:data
绝对路径:F:\program_work\java_work\CSDNCourse\FileProcessInCrawler\data\2.txt
相对路径:data\2.txt
文件的长度为:33
============
```

图 6.2　程序 6-2 的输出结果

6.1.3　文件字节流

1．InputStream 与 OutputStream

文件字节流用于读写字节。在网络爬虫处理图片、PDF、压缩和音频等文件时，可以使用文件字节流。但在使用文件字节流处理 Unicode 字符时，则会出现乱码现象，如读写包含汉字（一个汉字占用两字节）的文件。

在 java.io 包中，用于读入字节的相关类都继承（extends）了抽象类（abstract class）InputStream，用于写入字节的相关类都继承了抽象类 OutputStream。

在 InputStream 类中，提供了有关读字节及关闭流的方法，如表 6.3 所示。

表 6.3　InputStream 类中的有关读字节及关闭流的方法

方　　法	说　　明
abstract int read()	按顺序读入一字节，并返回该字节。如果到达源的末尾，则返回-1

续表

方　法	说　明
int read(byte b[])	读入一字节数组，返回读入的字节数。如果到达源的末尾，则返回-1。该方法最多读入 b.length 字节
int read(byte b[], int off, int len)	读入一字节数组，返回读入的字节数。off 表示首字节在数组 b 中的偏移量，len 表示读入字节的最大数量。如果到达源的末尾，则返回为-1
void close()	关闭输入流

在 OutputStream 类中，提供了有关写字节及关闭流的方法，如表 6.4 所示。

表 6.4　OutputStream 类中的有关写字节及关闭流的方法

方　法	说　明
abstract void write(int b)	写入一字节
void write(byte b[])	写入一字节数组
void write(byte b[], int off, int len)	从给定字节数组中起始于偏移量 off 处写 len 字节到文件
void close()	关闭输出流

由于抽象类 InputStream 和 OutputStream 只能读写单独的字节或字节数组，在实际应用中，我们使用的多为它们的子类。图 6.3 所示为 InputStream 的子类，图 6.4 所示为 OutputStream 的子类。

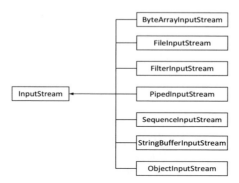

图 6.3　InputStream 的子类

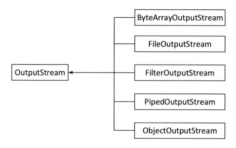

图 6.4　OutputStream 的子类

2. FileInputStream 与 FileOutputStream

FileInputStream 类是从 InputStream 类派生的简单输入流类，该类的相关操作方法继承自 InputStream 类。在使用 FileInputStream 时，需要创建 FileInputStream 对象，其创建对象的方法有两种，如表 6.5 所示。

表 6.5　FileInputStream 类中的创建对象的方法

方　　法	说　　明
FileInputStream(String name)	基于给定的文件名创建 FileInputStream 对象
FileInputStream(File file)	基于 File 对象创建 FileInputStream 对象

FileOutputStream 类是从 OutputStream 类派生的简单输出流类，该类的操作方法继承自 OutputStream 类。在使用 FileInputStream 时，需要创建 FileOutputStream 对象，其创建对象的方法也有两种，如表 6.6 所示。

表 6.6　FileOutputStream 类中的创建对象的方法

方　　法	说　　明
FileOutputStream(String name)	基于给定的文件名创建 FileOutputStream 对象
FileOutputStream(File file)	基于 File 对象创建 FileOutputStream 对象

程序 6-3 所示为 FileInputStream 与 FileOutputStream 的使用情况。

程序 6-3

```java
import java.io.File;
import java.io.FileInputStream;
import java.io.FileOutputStream;
import java.io.IOException;
public class StreamTest1 {
  public static void main(String[] args) throws IOException {
      // FileInputStream的两种构造方法
      //FileInputStream inputStream = new FileInputStream("data/1.txt");
      FileInputStream inputStream = new FileInputStream(new File("data/1.txt"));
      // FileOutputStream的两种构造方法
      FileOutputStream outputStream = new FileOutputStream("data/out.txt");
      //FileOutputStream outputStream = new FileOutputStream(new File("data/out.txt"));
      int temp;
      //按照字节进行读写
```

```
            while ((temp = inputStream.read()) != -1) {
                System.out.print((char)temp);     //输出结果
                outputStream.write(temp);          //写入指定文件
            }
            //流的关闭
            outputStream.close();
            inputStream.close();
        }
    }
```

执行程序 6-3，会在控制台输出 "data/1.txt" 文件的文本内容，如下所示。同时，也会发现，在 "data/" 目录下，生成了一个新的文本文件 "out.txt"，并且这个新的文本文件内容与 "1.txt" 的一样。

6.1.4　文件字符流

1. Reader 与 Writer

文件字符流主要用于读写字符，每字符可以占用多字节。例如，在处理含有中文字符的文本时，便可以使用文件字符流。在 java.io 包中，读入字符的相关类都继承了抽象类 Reader，写入字符的相关类都继承了抽象类 Writer。

Reader 类中的方法如表 6.7 所示，可以看到这些方法与 InputStream 类中的方法很类似。

表 6.7　Reader 类中的方法

方　　法	说　　明
int read()	读取一字符，返回 Unicode 字符值（0~65535 的整数），如果到达源的末尾，则返回-1
int read(char cbuf[])	最多读取 b.length 字符到数组中，如果到达源的末尾，则返回-1
int read(char cbuf[], int off, int len)	读取 len 字符并存放到数组中，off 为首字符的数组中的位置，返回实际读取字符的数目。如果到达文件末尾，则返回-1
int read(java.nio.CharBuffer target)	将字符先添加到缓冲区，然后读取
void close()	关闭流

Writer 类中的方法如表 6.8 所示。

表 6.8　Writer 类中的方法

方　　法	说　　明
void write(int c)	向文件写入一字符
void write(char cbuf[])	向文件写入一个字符数组

续表

方　　法	说　　明
void write(char cbuf[], int off, int len)	从字符数组中起始偏移量 off 处读取 len 字符到文件
void write(String str)	向文件写入字符串
void write(String str, int off, int len)	从字符串中起始偏移量 off 处读取 len 字符到文件
Writer append(char c)	向文件添加单字符
Writer append(CharSequence csq)	向文件中添加字符序列
Writer append(CharSequence csq, int start, int end)	从字符序列中起始偏移量 off 处读取 len 字符到文件
void close()	关闭流

在实际应用中，主要使用的是 Reader 与 Writer 的子类。图 6.5 所示为 Reader 的子类，图 6.6 所示为 Writer 的子类。

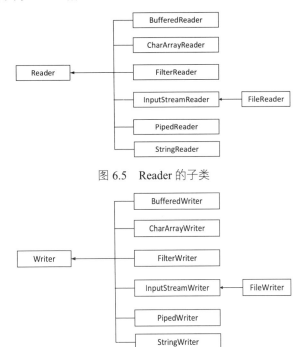

图 6.5　Reader 的子类

图 6.6　Writer 的子类

2. InputStreamReader 与 OutputStreamWriter

InputStreamReader 类是字符流 Reader 的子类，是字节流通向字符流的桥梁。在使用 InputStreamReader 类中的方法创建 InputStreamReader 对象时，可以指定编码方法，如"utf-8"，如果不指定编码，则采用默认编码"gbk"。表 6.9 所示为 InputStreamReader 类中的方法。

表 6.9　InputStreamReader 类中的方法

方　法	说　明
InputStreamReader(InputStream in)	基于给定的 InputStream 对象创建 InputStreamReader 对象
InputStreamReader(InputStream in, String charsetName)	基于给定的 InputStream 对象以及指定的编码方式（String 类型）创建 InputStreamReader 对象
InputStreamReader(InputStream in, Charset cs)	基于给定的 InputStream 对象以及指定的编码方式（Charset 类型）创建 InputStreamReader 对象
InputStreamReader(InputStream in, CharsetDecoder dec)	基于给定的 InputStream 对象以及指定的编码方式（CharsetDecoder 类型）创建 InputStreamReader 对象

OutputStreamWriter 是字符流 Writer 的子类，是字符流通向字节流的桥梁，OutputStreamWriter 类中的方法如表 6.10 所示。

表 6.10　OutputStreamWriter 类中的方法

方　法	说　明
OutputStreamWriter(OutputStream out)	基于给定的 OutputStream 对象创建 OutputStreamWriter 对象
OutputStreamWriter(OutputStream out, String charsetName)	基于给定的 OutputStream 对象以及指定的编码方式（String 类型）创建 OutputStreamWriter 对象
OutputStreamWriter(OutputStream out, Charset cs)	基于给定的 OutputStream 对象以及指定的编码方式（Charset 类型）创建 OutputStreamWriter 对象
OutputStreamWriter(OutputStream out, CharsetEncoder enc)	基于给定的 OutputStream 对象以及指定的编码方式（CharsetDecoder 类型）创建 OutputStreamWriter 对象

程序 6-4 所示为 InputStreamReader 与 OutputStreamWriter 的使用情况。

程序 6-4

```java
import java.io.FileInputStream;
import java.io.FileOutputStream;
import java.io.IOException;
import java.io.InputStream;
import java.io.InputStreamReader;
import java.io.OutputStreamWriter;
public class IrOwTest {
 public static void main(String[] args) throws IOException {
    //读取文件中的数据
    InputStream in = new FileInputStream("data/3.txt");
    //字节流向字符流的转换
    InputStreamReader isr = new InputStreamReader(in);
```

```java
    //写入新文件
    FileOutputStream fos = new FileOutputStream("data/3osw.txt");
    OutputStreamWriter osw = new OutputStreamWriter(fos,"utf-8");
    int temp;
    //读取与写入操作
    while((temp = isr.read()) != -1){
        System.out.print((char)temp);
        osw.write(temp);
    }
    osw.write("\njava网络爬虫");
    osw.append("\n很有意思");
    //流的关闭
    osw.close();
    isr.close();
 }
}
```

执行该程序,会输出"3.txt"文件中的文本内容,同时会在 data 目录下创建一个新的"3osw.txt",并且写入新的文本内容。

3. FileReader 与 FileWriter

FileReader 与 FileWriter 分别继承了 InputStreamReader 与 OutputStreamWriter。表 6.11 所示为 FileReader 类中的方法。

表 6.11 FileReader 类中的方法

方 法	说 明
FileReader(String fileName)	基于文件名创建 FileReader 对象
FileReader(File file)	基于 File 对象创建 FileReader 对象

表 6.12 所示为 FileWriter 类中的方法。

表 6.12 FileWriter 类中的方法

方 法	说 明
FileWriter(String fileName)	基于文件名创建 FileWriter 对象
FileWriter(File file)	基于 File 对象创建 FileWriter 对象
FileWriter(String fileName, boolean append)	基于文件名创建 FileWriter 对象,append 表示在原有文件内容的基础上追加内容
FileWriter(File file, boolean append)	基于 File 对象创建 FileWriter 对象,append 表示在原有文件内容的基础上追加内容

程序 6-5 所示为 FileReader 与 FileWriter 的使用情况。

程序 6-5

```java
import java.io.FileReader;
import java.io.FileWriter;
import java.io.IOException;
public class FrFWTest {
    public static void main(String[] args) throws IOException {
        //两种文件字符输入流创建方式
        FileReader fileReader = new FileReader("data/3.txt");
        //FileReader fileReader = new FileReader(new File("data/1.txt"));
        //两种文件字符输出流创建方式
        FileWriter fileWriter = new FileWriter("data/outtest.txt");
        //FileWriter fileWriter = new FileWriter(new File("data/outtest.txt",true));
        int temp;
        while ((temp = fileReader.read()) != -1) {
            System.out.print((char)temp);
            fileWriter.write((char)temp);
        }

        fileWriter.write("我爱网络爬虫", 2, 3);
        //StringBuilder实现了CharSequence
        fileWriter.append(new StringBuilder("\n轻松学习java"));
        fileWriter.close();
        fileReader.close();
    }
}
```

6.1.5 缓冲流

1. 简介

普通的字节流与字符流都是无缓冲的输入和输出流，操作单元是字节或字符，每次读写都涉及对磁盘的操作，效率较低。为增强文件的读写能力，java.io 包提供了更高级的流——缓冲流。缓冲流包括四个类：BufferedInputStream 类、BufferedOutputStream 类、BufferedReader 类和 BufferedWriter 类，如图 6.7 所示。

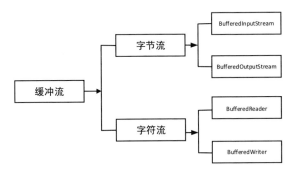

图 6.7　缓冲流中的字节流与字符流

2．BufferedInputStream 和 BufferedOutputStream

BufferedInputStream 和 BufferedOutputStream 为缓冲流中的字节流，它们分别是 FilterInputStream 和 FilterOutputStream 的子类。网络爬虫在采集图片、PDF、视频和压缩文件时，可优先考虑使用 BufferedInputStream 和 BufferedOutputStream。例如，在 4.1.7 节和 4.1.9 节中，使用 Jsoup 下载图片及 tar.gz 压缩文件皆使用了这两种缓冲流。表 6.13 所示为 BufferedInputStream 类中的方法，表 6.14 所示为 BufferedOutputStream 类中的方法。

表 6.13　BufferedInputStream 类中的方法

方　　法	说　　明
BufferedInputStream(InputStream in)	创建默认大小缓冲区的缓冲字节输入流
BufferedInputStream(InputStream in, int size)	创建指定大小缓冲区的缓冲字节输入流

表 6.14　BufferedOutputStream 类中的方法

方　　法	说　　明
BufferedOutputStream(OutputStream out)	创建默认大小缓冲区的缓冲字节输出流
BufferedOutputStream(OutputStream out, int size)	创建指定大小缓冲区的缓冲字节输出流

程序 6-6 所示为 BufferedInputStream 和 BufferedOutputStream 的使用情况。

程序 6-6

```java
import java.io.BufferedInputStream;
import java.io.BufferedOutputStream;
import java.io.File;
import java.io.FileInputStream;
import java.io.FileOutputStream;
import java.io.IOException;
public class BisBos {
 public static void main(String[] args) throws IOException {
```

```
//创建File对象
File file = new File("data/1.txt");
//创建BufferedInputStream对象
BufferedInputStream bin = new BufferedInputStream(
        new FileInputStream(file), 512);
//创建BufferedOutputStream对象
BufferedOutputStream bos = new BufferedOutputStream(
        new FileOutputStream("data/1bos.txt"));
byte[] b=new byte[5];   //代表一次最多读取5字节的内容
int length = 0;         //代表实际读取的字节数
while( (length = bin.read( b ) )!= -1 ){
    System.out.println(new String(b));
    //写入指定文件
    bos.write(b, 0, length);
    bos.write("\n".getBytes()); //写上换行符
}
//关闭流
bos.close();
bin.close();
    }
}
```

在程序 6-6 中，设置了每次最多读取 5 字节的内容，因此在控制台会输出多段结果，每段结果最多包含 5 个英文字符，如图 6.8 所示。

图 6.8 程序 6-6 的输出结果

3．BufferedReader 和 BufferedWriter

BufferedReader 和 BufferedWriter 为缓冲流中的字符流。由图 6.4 和图 6.5 可以发现这两个类分别是 Reader 和 Writer 的子类。在 BufferedReader 类中，提供了按行读取字符串的方法，其用来处理文本非常方便。在网络爬虫以及自然语言处理算法中，涉及文本数据读取、存储以及预处理的部分，基本都要用到 BufferedReader 和 BufferedWriter 这两个类。表 6.15 所示为 BufferedReader 类中的方法，表 6.16 所示为 BufferedWriter 类中的方法。

表 6.15　BufferedReader 类中的方法

方　　法	说　　明
BufferedReader(Reader in)	创建默认大小缓冲区的缓冲字符输入流
BufferedReader(Reader in, int sz)	创建指定大小缓冲区的缓冲字符输入流

表 6.16　BufferedWriter 类中的方法

方　　法	说　　明
BufferedWriter(Writer out)	创建默认大小缓冲区的缓冲字符输出流
BufferedWriter(Writer out, int sz)	创建指定大小缓冲区的缓冲字符输入流

程序 6-7 所示为 BufferedReader 和 BufferedWriter 的使用情况，包括流对象的创建、按行读取文件和按行写入文件、Map 集合数据写入指定文件。

程序 6-7

```java
import java.io.BufferedReader;
import java.io.BufferedWriter;
import java.io.File;
import java.io.FileInputStream;
import java.io.FileOutputStream;
import java.io.FileReader;
import java.io.IOException;
import java.io.InputStreamReader;
import java.io.OutputStreamWriter;
import java.util.HashMap;
import java.util.Map;
public class BufferedTest {
 public static void main(String[] args) throws IOException {
    /****** 文件读取第一种方式　******/
    File file = new File("data/3.txt");
    //FileReader读取文件
    FileReader fileReader = new FileReader(file);
    //根据FileReader创建缓冲流
    BufferedReader bufferedReader = new BufferedReader(fileReader);
    String s = null;
    //按行读取
    while ((s = bufferedReader.readLine())!=null) {
        System.out.println(s);
    }
    //流关闭
    bufferedReader.close();
    fileReader.close();
```

```java
        /****** 文件读取第二种方式  ******/
        //这里进行了简写，写成了一行。可以添加字符编码
        BufferedReader reader = new BufferedReader( new
InputStreamReader(
                new FileInputStream(
                        new File( "data/3.txt")),"utf-8"));
        String s1 = null;
        while ((s1 = reader.readLine())!=null) {
            System.out.println(s1);
        }
        //流关闭
        reader.close();
        /****** 文件写入第一种方式  ******/
        /*File file1 = new File("data/bufferedout.txt","gbk");
        FileOutputStream fileOutputStream = new FileOutputStream
(file1);
        OutputStreamWriter outputStreamWriter = new OutputStreamWriter
(fileOutputStream);
        BufferedWriter bufferedWriter1 = new BufferedWriter
(outputStreamWriter);*/
        /****** 文件写入快捷方式******/
        BufferedWriter writer = new BufferedWriter( new
OutputStreamWriter
                ( new FileOutputStream(
                        new File("data/bufferedout.txt")),"gbk"));
        Map<Integer,String> map = new HashMap<Integer,String>();
        map.put(0, "http://pic.******.com/list/2_0_2.html");
        map.put(1, "http://pic.******.com/list/2_0_3.html");
        map.put(2, "http://pic.******.com/list/2_0_4.html");
        //Map遍历数据
        for( Integer key : map.keySet() ){
            writer.append("key:"+key+"\tvalue:"+map.get(key));
            writer.newLine();  //写入换行操作
        }
        //流关闭
        writer.close();
    }
}
```

6.1.6 网络爬虫下载图片实战

在前面已经介绍，下载图片、PDF和压缩文件时，可优先考虑缓冲流中的字节流。

下面，将通过一个网络爬虫案例进行讲解，所要采集的数据为vecteezy图库（见图6.9）中的图片。

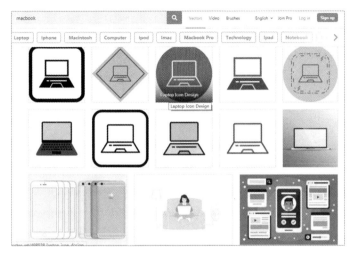

图 6.9　vecteezy 图库

首先，可以使用HttpClient请求该网址，并解析得到该页面中的每张图片的URL。同时，基于HttpClient请求每张图片的URL，获取实体内容（HttpEntity）。最后，使用BufferedOutputStream将图片写入指定目录文件。完整的实现代码如程序6-8所示。

程序 6-8

```java
package com.crawler;

import java.io.BufferedInputStream;
import java.io.BufferedOutputStream;
import java.io.File;
import java.io.FileOutputStream;
import java.io.IOException;
import java.io.InputStream;
import org.apache.http.HttpEntity;
import org.apache.http.HttpResponse;
import org.apache.http.client.ClientProtocolException;
import org.apache.http.client.HttpClient;
import org.apache.http.client.methods.HttpGet;
import org.apache.http.impl.client.HttpClients;
import org.apache.http.util.EntityUtils;
import org.jsoup.Jsoup;
import org.jsoup.nodes.Element;
import org.jsoup.select.Elements;
```

```java
public class CrawPictureVec {
    private static HttpClient httpClient = HttpClients.custom().build();
    public static void main(String[] args) throws IOException{
            String url = "https://www.********.com/"
                + "free-vector/macbook?page=1"
                + "&from=mainsite&in_se=true";
        HttpEntity entity = getEntityByHttpGetMethod(url);
        //获取所有图片链接
        String html = EntityUtils.toString(entity);
        Elements elements = Jsoup.parse(html).select("#main > ul > li > a > img");
        for (Element ele : elements) {
            String pictureUrl = ele.attr("data-lazy-src");
            saveImage(pictureUrl,"image/" + pictureUrl.split("/")[10]);
        }
        //测试程序
        saveImage1("https://static.********."
                + "com/system/resources/thumbnails"
                + "/000/498/574/small/"
                + "Education_31-60_1054.jpg","image/1.jpg");
    }
    //请求某一个URL，获得请求到的内容
    public static HttpEntity getEntityByHttpGetMethod(String url){
        HttpGet httpGet = new HttpGet(url);
        //获取结果
        HttpResponse httpResponse = null;
        try {
            httpResponse = httpClient.execute(httpGet);
        } catch (IOException e) {
            e.printStackTrace();
        }
        HttpEntity entity = httpResponse.getEntity();
        return entity;
    }
    //输入任意地址便可以下载图片
    static void saveImage(String url, String savePath) throws IOException{
        //图片下载保存地址
        File file=new File(savePath);
        //如果文件存在则删除
```

```java
    if(file.exists()){
        file.delete();
    }
    //缓冲流
    BufferedOutputStream bw = new BufferedOutputStream(
            new FileOutputStream(savePath));
    //请求图片数据
    try {
        HttpEntity entity = getEntityByHttpGetMethod(url);
        //以字节的方式写入
        byte[] byt= EntityUtils.toByteArray(entity);
        bw.write(byt);
        System.out.println("图片下载成功! ");
    } catch (ClientProtocolException e) {
        e.printStackTrace();
    } catch (IOException e) {
        e.printStackTrace();
    }
    //关闭缓冲流
    bw.close();
}

//另一种操作方式
static void saveImage1(String url, String savePath)
        throws UnsupportedOperationException, IOException {
    //获取图片信息，作为输入流
    InputStream in = getEntityByHttpGetMethod(url).getContent();
    //每次最多读取1KB的内容
    byte[] buffer = new byte[1024];
    BufferedInputStream bufferedIn =
            new BufferedInputStream(in);
    int len = 0;
    //创建缓冲流
    FileOutputStream fileOutStream = new FileOutputStream(new File(savePath));
    BufferedOutputStream bufferedOut = new BufferedOutputStream(fileOutStream);
    //图片写入
    while ((len = bufferedIn.read(buffer, 0, 1024)) != -1) {
```

```
            bufferedOut.write(buffer, 0, len);
        }
        //缓冲流释放与关闭
        bufferedOut.flush();
        bufferedOut.close();
    }
}
```

在程序 6-8 中，getEntityByHttpGetMethod(String url)方法的作用是请求某个具体 URL，获取实体 HttpEntity。同时，程序 6-8 中提供了两种方法实现图片下载功能，即 saveImage(String url, String savePath)和 saveImage1(String url, String savePath)方法，其本质都是对字节数据的操作。在程序 6-8 的 main 方法中，使用了 Jsoup 解析 HTML 页面获取每张图片的 URL，调用 saveImage(String url, String savePath)方法便可成功下载每张图片。图 6.10 所示为程序 6-8 下载的部分图片。

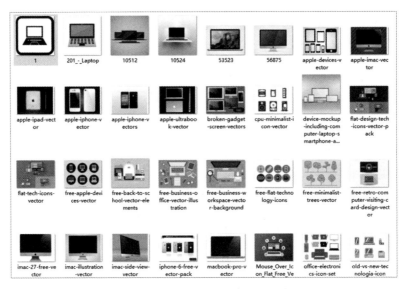

图 6.10　程序 6-8 下载的部分图片

6.1.7　网络爬虫文本存储实战

下面通过一个案例，介绍网络爬虫中的文本操作。需要采集的是 pythonforbeginners 网站（见图 6.11）中的前三页的帖子信息，涉及的 URL 列表如下所示。

```
https://www.******************.com/?page=1
https://www.******************.com/?page=2
https://www.******************.com/?page=3
```

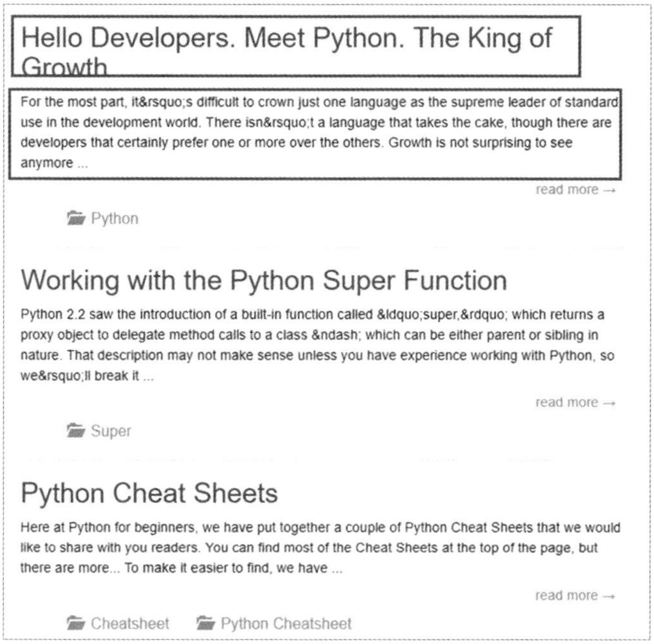

图 6.11　pythonforbeginners 网站中的帖子

在编写程序之前，首先确定需要采集的字段，如这里需要采集的字段是帖子标题和帖子简介。接着，基于需要采集的字段，创建 PostModel 类（JavaBean），如程序 6-9 所示。

程序 6-9

```java
package com.crawler.pythonforbeginners;

public class PostModel {
 private String post_title;      //帖子标题
 private String post_content;    //帖子简介
 public String getPost_title() {
     return post_title;
 }
 public void setPost_title(String post_title) {
     this.post_title = post_title;
 }
 public String getPost_content() {
     return post_content;
 }
 public void setPost_content(String post_content) {
     this.post_content = post_content;
```

    }

  }

下一步，确定需要使用的 URL 请求工具，这里使用较为简单的 Jsoup。同时，需要在浏览器中定位所抽取字段对应的 HTML 标签（也可通过网络抓包），如图 6.12 所示。

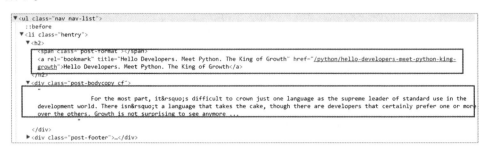

图 6.12　所抽取字段在网页中的位置

最后，编写获取 HTML 数据（Jsoup）、解析数据（Jsoup）和保存数据（BufferedWriter）的程序，如程序 6-10 所示。

程序 6-10

```java
package com.crawler.pythonforbeginners;

import java.io.BufferedWriter;
import java.io.File;
import java.io.FileOutputStream;
import java.io.IOException;
import java.io.OutputStreamWriter;
import java.util.ArrayList;
import java.util.List;
import org.jsoup.Jsoup;
import org.jsoup.nodes.Document;
import org.jsoup.nodes.Element;
import org.jsoup.select.Elements;


public class CrawlerTest {

  public static void main(String[] args) throws IOException {
    //待爬URL列表
```

```java
        List<String> urlList = new ArrayList<String>();
        urlList.add("https://www.*****************.com/?page=1");
        urlList.add("https://www.*****************.com/?page=2");
        urlList.add("https://www.*****************.com/?page=3");
        //缓冲流的创建，以gbk写入文本
        BufferedWriter writer = new BufferedWriter(
                new OutputStreamWriter(
                        new FileOutputStream(
                            new File("crawldata/"
                            + "pythonforbeginners.txt")),"gbk"));
        for (int i = 0; i < urlList.size(); i++) {
            List<PostModel> data = crawerData(urlList.get(i));
            for (PostModel model : data) {
                //所爬数据写入文本
                writer.write(model.getPost_title() + "\t" + model.
getPost_content() + "\r\n");
            }
        }

        //流的关闭
        writer.close();
    }
    static List<PostModel> crawerData(String url) throws IOException{
        //所爬数据封装于集合中
        List<PostModel> datalist = new ArrayList<PostModel>();
        //获取URL对应的HTML内容
        Document doc = Jsoup.connect(url).timeout(30000).get();
        //定位需采集的每个帖子
        Elements elements = doc.select("ul[class=nav nav-list]")
                .select("li[class=hentry]");
        //遍历每一个帖子
        for (Element ele : elements) {
            //解析数据
            String post_title = ele.select(" h2 > a").text();
            String post_content = ele
                    .select("div[class=post-bodycopy cf]").text();
            //创建对象和封装数据
            PostModel model = new PostModel();
            model.setPost_title(post_title);
            model.setPost_content(post_content);
            datalist.add(model);
```

```
        }
        return datalist;
    }
}
```

执行程序 6-10，可以发现在程序的 "crawldata/" 目录下成功创建了 "pythonforbeginners.txt" 文本文件，并且文本文件中的内容为所爬的帖子标题和帖子简介，如图 6.13 所示。

```
1 Hello Developers. Meet Python. The King of Growth   For the most part, it&rsquo;s difficult t
2 Working with the Python Super Function   Python 2.2 saw the introduction of a built-in functio
3 Python Cheat Sheets Here at Python for beginners, we have put together a couple of Python Che
4 Beautiful Soup 4 Python Overview This article is an introduction to BeautifulSoup 4 in Python
5 Web Scraping with BeautifulSoup Web Scraping "Web scraping (web harvesting or web data extrac
6 Python - Quick Guide    "Python Basics We have updated our "Python - Quick Guide". You can fin
7 The del Statement    The del statement can be used to remove an item from a list by referring
8 __str__ vs. __repr__    According to the official Python documentation, __repr__ is a built-i
9 Break and Continue Statements   Break statements exist in Python to exit or "break" a for or
10 Numeric Types in Python In any OOP language, there are many different data types. In Python,
11 DNS Lookup With Python   The sockets module provides an easy way to look up a host name's ip a
12 Encoding JSON with Python   Python comes pre-equipped with a JSON encoder and decoder to make
13 How to Use MySQL Connector/Python    MySQL Connector/Python is a driver released by Oracle fo
14 Getting the most popular pages from your Apache logfile An Apache logfile can be huge and har
15 Make your life easier with Virtualenvwrapper   When you do a lot of Python programming, you
16 This site now runs on Django    Today we finished migrating the old PythonForBeginners.com fr
17 How to use Pillow, a fork of PIL   Overview In last post I was writing about PIL, also known
18 How to use the Python Imaging Library    PIL is deprecated, obsolete. please use Pillow. Find
19 Python Websites and Tutorials    The list below is made to help new Python programmers to find
20 How to use Envoy    About Envoy Recently I stumble upon Envoy. Envoy is a wrapper around the
21 Using Feedparser in Python   Overview In this post we will take a look on how we can download
22 Subprocess and Shell Commands in Python Subprocess Overview For a long time I have been using
```

图 6.13 "pythonforbeginners.txt" 文本文件中的内容

6.2　Excel 存储

网络爬虫采集少量数据时，可以使用 Excel 进行存储。Java 中主要的两款操作 Excel 的工具是 Jxl 和 Apache 旗下的 POI。

6.2.1　Jxl 的使用

Jxl 是一款常用的 Java 中操作 Excel 的 API，但其只对 xls 有效，对 2007 版本以上的 Excel（xlsx）很难处理。在本节中，主要介绍如何使用 Jxl 创建工作簿以及工作表、读取 Excel 文件内容和写入 Excel 文件内容。关于其他方面的操作（如调整 Excel 的宽高和颜色设置等），读者可自行学习。

1. jar 包的下载

首先，基于 Maven 工程的 pom.xml 文件配置 Jxl。

```
        <!-- https://*************.com/artifact/net.sourceforge.jexcelapi/jxl -->
        <dependency>
```

```
<groupId>net.sourceforge.jexcelapi</groupId>
<artifactId>jxl</artifactId>
<version>2.6.12</version>
</dependency>
```

2. 数据写入

在写入 Excel 之前，需要创建 Excel 工作簿以及工作表，如下语句。

```java
File xlsFile = new File("data/a.xls");
// 创建一个工作簿
WritableWorkbook workbook = Workbook.createWorkbook(xlsFile);
// 创建一个工作表
WritableSheet sheet = workbook.createSheet("sheet1", 0);
```

创建完工作表之后，可以使用 addCell(WritableCell cell)方法向数据表中添加数据，如下为 Excel 工作表添加一个表头。

```java
//添加表头
sheet.addCell(new Label(0, 0, "post_id"));
sheet.addCell(new Label(1, 0, "post_title"));
```

其中，构造方法 Label(int c, int r, String cont)中的 c 表示列，r 表示行，cont 表示文本内容。如上面的操作表示向第一行的第一列和第二列分别添加字符"post_id"和"post_title"。添加完表头内容后，可以使用循环语句，继续向 Excel 工作表中添加数据。

```java
//添加内容
for(int i = 0; i < 2; i++){
    sheet.addCell(new Label(0, i+1, "0" + i));
    sheet.addCell(new Label(1, i+1, "内容" + i));
}
```

之后，执行写入文件操作，关闭资源，释放内存。

```java
//执行写入文件操作
workbook.write();
//关闭资源，释放内存
workbook.close();
```

通过上述一系列的操作，可以成功将表头和数据写入指定的 Excel 工作簿中，如图 6.14 所示。

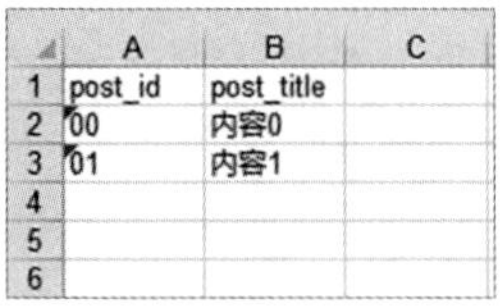

	A	B	C
1	post_id	post_title	
2	00	内容0	
3	01	内容1	
4			
5			
6			

图 6.14　Excel 工作簿中的内容

3. 数据读取

使用 Jxl 读取 Excel 文件（xls）需要声明工作簿，即确定读取的 Excel 文件。Workbook 类中提供了多种方法，声明读取的工作簿，如下所示。

```
Workbook getWorkbook(java.io.File file)      //基于 File 对象
Workbook getWorkbook(java.io.File file, WorkbookSettings ws)
Workbook getWorkbook(InputStream is)         //基于字节流
Workbook getWorkbook(InputStream is, WorkbookSettings ws)
```

声明工作簿后，需要确定读取的工作表，Workbook 类中提供了两种方法确定操作的工作表。

```
Sheet getSheet(String name)       //基于 sheet 名
Sheet getSheet(int index)         //基于 sheet 索引
```

接着，获取工作表的所有行数和列数，循环读出所有数据。程序 6-11 给出了读取 Excel 的完整代码。

程序 6-11

```
import java.io.File;
import java.io.IOException;
import jxl.Cell;
import jxl.Sheet;
import jxl.Workbook;
import jxl.read.biff.BiffException;
import jxl.write.WriteException;
import jxl.write.biff.RowsExceededException;
public class JxlExcelRead {
  public static void main(String[] args) throws IOException,
RowsExceededException, WriteException, BiffException {
        // 声明工作簿
        Workbook book  = Workbook.getWorkbook(new File("data/a.xls"));
        //获取名称为sheet1的表格，也可使用getSheet(0)获取第一个工作表
        Sheet sheet = book.getSheet("sheet1");
        //获取工作表中的总行数及总列数
```

```java
        int rows = sheet.getRows();
        int columns = sheet.getColumns();
        //行列循环获取数据
        for (int i = 0; i < rows; i++) {
           for (int j = 0; j < columns; j++) {
               //使用getCell方法读取数据,
               //第一个参数是指定第几列,第二个参数是指定第几行
               Cell cell= sheet.getCell(j,i);
               System.out.print(cell.getContents() + "\t");
           }
           System.out.println();
        }
   }
}
```

图 6.15 所示为程序 6-11 在控制台的输出结果。

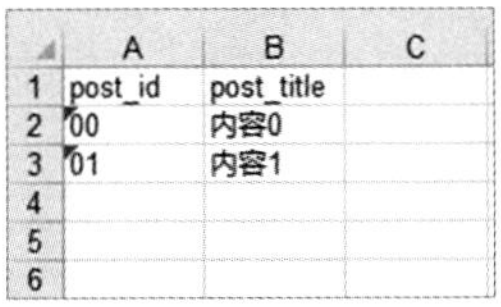

图 6.15　程序 6-11 的输出结果

6.2.2　POI 的使用

POI 是 Java 编写的开源跨平台 Excel 处理工具,功能比 Jxl 更强大。POI 不仅提供了对 Excel 的操作,也提供了对 Word、PowerPoint 和 Visio 等格式的文档的操作。

1. jar 包的下载

首先,基于 Maven 工程的 pom.xml 文件配置 POI。

```xml
        <!-- https://*************.com/artifact/org.apache.poi/poi -->
        <dependency>
            <groupId>org.apache.poi</groupId>
            <artifactId>poi</artifactId>
            <version>3.17</version>
        </dependency>
        <!-- https://*************.com/artifact/org.apache.poi/poi-ooxml -->
        <dependency>
```

```
            <groupId>org.apache.poi</groupId>
            <artifactId>poi-ooxml</artifactId>
            <version>3.17</version>
</dependency>
```

2. 相关类介绍

POI 既可以操作 xls 文件（2007 版以前的 Excel）也可以操作 xlsx 文件（2007 版以后的 Excel）。表 6.17 所示为处理 xls 常用的类，表 6.18 所示为处理 xlsx 文件常用的类。

表 6.17　处理 xls 常用的类

类	说　　明
HSSFWorkbook	Excel 工作簿 Workbook
HSSFSheet	Excel 工作表 Sheet
HSSFRow	Excel 行
HSSFCell	Excel 单元格

表 6.18　处理 xlsx 常用的类

类	说　　明
XSSFWorkbook	Excel 工作簿 Workbook
XSSFSheet	Excel 工作表 Sheet
XSSFRow	Excel 行
XSSFCell	Excel 单元格

从 POI 的源码中可以发现处理 xls 文件的类与处理 xlsx 文件的类存在如表 6.19 所示的关系。

表 6.19　实现的接口

类	实现的接口
HSSFWorkbook 和 XSSFWorkbook	Workbook 接口
HSSFSheet 和 XSSFSheet	Sheet 接口
HSSFRow 和 XSSFRow	Row 接口
HSSFCell 和 XSSFCell	Cell 接口

基于对以上类的了解，下面开始介绍如何使用 POI 进行 Excel 数据的读写，其中涉及 xls 和 xlsx 两种格式。

3. 数据写入

针对 xls 格式的 Excel 数据，需要使用 HSSF 开头的类进行操作。其中，数据写入 Excel 可按如下步骤进行。

步骤 1：创建文件输入流，确定要写入的文件目录。

```
//创建文件输出流
File file = new File("data/b.xls");
OutputStream outputStream = new FileOutputStream(file);
```

步骤 2：使用 HSSFWorkbook 类创建工作簿、使用 HSSFSheet 类创建工作表。

```
//创建工作簿及工作表
HSSFWorkbook workbook = new HSSFWorkbook();
HSSFSheet sheet = workbook.createSheet("Sheet1");
```

步骤 3：使用 HSSFSheet 类中的 createRow(int rownum)方法创建某行，使用 HSSFRow 类中的 createCell(int column)方法创建某列，使用 HSSFCell 类中的 setCellValue(String value)方法为该单元格赋值，代码如下所示。

```
//添加表头
HSSFRow row = sheet.createRow(0); //常见某行
row.createCell(0).setCellValue("post_id");
row.createCell(1).setCellValue("post_title");
//添加内容
for(int i = 0; i < 2; i++){
    HSSFRow everyRow = sheet.createRow(i + 1);
    everyRow.createCell(0).setCellValue("帖子id为：0" + i);
    everyRow.createCell(1).setCellValue("帖子内容为：" + i);
}
```

步骤 4：数据写入工作目录，并释放资源。

```
workbook.write(outputStream);
workbook.close();
outputStream.close();
```

基于上述 4 个步骤，可将数据成功写入 xls 格式的 Excel，如图 6.16 所示。

图 6.16　POI 写入 xls 格式的 Excel

使用 POI 将数据写入 xlsx 格式的 Excel，流程与写入 xls 格式的 Excel 相同，唯一的区别是将 HSSF 开头的类，换成 XSSF 开头的类，程序 6-12 给出了完整的代码。

程序 6-12

```java
import java.io.File;
import java.io.FileOutputStream;
import java.io.IOException;
import java.io.OutputStream;
import org.apache.poi.xssf.usermodel.XSSFRow;
import org.apache.poi.xssf.usermodel.XSSFSheet;
import org.apache.poi.xssf.usermodel.XSSFWorkbook;
public class PoiExcelWriteXlsx {
    public static void main(String[] args) throws IOException {
        //创建文件输出流
        File file = new File("data/b.xlsx");
        OutputStream outputStream = new FileOutputStream(file);
        //创建工作簿及工作表
        XSSFWorkbook workbook = new XSSFWorkbook();
        XSSFSheet sheet = workbook.createSheet("Sheet1");
        //添加表头
        XSSFRow row = sheet.createRow(0); //创建某行
        row.createCell(0).setCellValue("post_id");
        row.createCell(1).setCellValue("post_title");
        //添加内容
        for(int i = 0; i < 2; i++){
            XSSFRow everyRow = sheet.createRow(i + 1);
            everyRow.createCell(0).setCellValue("帖子id为: 0" + i);
            everyRow.createCell(1).setCellValue("帖子内容为: " + i);
        }
        //数据写入工作目录，并释放资源
        workbook.write(outputStream);
        workbook.close();
        outputStream.close();
    }
}
```

执行程序 6-12，发现在"data"目录下，成功创建了"b.xlsx"文件，且表格中写入了 3 行数据。

另外，从表 6.17 中可以发现，处理 xls 格式的文件的类与处理 xlsx 格式的文件的类实现的接口相同。基于这种关系，程序 6-13 给出了既可以将数据写入 xls 格式的文件也可以写入 xlsx 文件的代码。

程序 6-13

```java
import java.io.File;
import java.io.FileOutputStream;
import java.io.IOException;
import java.io.OutputStream;
import org.apache.poi.hssf.usermodel.HSSFWorkbook;
import org.apache.poi.ss.usermodel.Row;
import org.apache.poi.ss.usermodel.Sheet;
import org.apache.poi.ss.usermodel.Workbook;
import org.apache.poi.xssf.usermodel.XSSFWorkbook;
public class PoiExceWritelProcess {
    public static void main(String[] args) throws IOException {
        //文件名称
        File file = new File("data/c.xls");
        //File file = new File("data/c.xlsx");
        OutputStream outputStream = new FileOutputStream(file);
        Workbook workbook = getWorkBook(file);
        Sheet sheet = workbook.createSheet("Sheet1");
        //添加表头
        Row row = sheet.createRow(0); //创建某行
        row.createCell(0).setCellValue("post_id");
        row.createCell(1).setCellValue("post_title");
        //添加内容
        for(int i = 0; i < 2; i++){
            Row everyRow = sheet.createRow(i + 1);
            everyRow.createCell(0).setCellValue("帖子id为: 0" + i);
            everyRow.createCell(1).setCellValue("帖子内容为: " + i);
        }
        workbook.write(outputStream);
        //释放资源
        workbook.close();
        outputStream.close();
    }
    /**
     * 判断Excel的版本，初始化不同的Workbook
     * @param filename
     * @return
     * @throws IOException
     */
    public static Workbook getWorkBook(File file) throws IOException{
        Workbook workbook = null;
        //Excel 2003
```

```java
    if(file.getName().endsWith("xls")){
        workbook = new HSSFWorkbook();
    // Excel 2007以上版本
    }else if(file.getName().endsWith("xlsx")){
        workbook = new XSSFWorkbook();
    }
    return workbook;
 }
}
```

4. 数据读取

POI 读取 xls 格式或 xlsx 格式的数据与其写入数据的方式类似，可以单独使用各自的类及类中的方法，读取表格中的数据，也可先判断 Excel 格式是 xls 还是 xlsx。程序 6-14 给出了既可以读取 xls 格式的文件也可以读取 xlsx 格式的文件的代码。

程序 6-14

```java
import java.io.File;
import java.io.FileInputStream;
import java.io.IOException;
import java.io.InputStream;
import org.apache.poi.hssf.usermodel.HSSFWorkbook;
import org.apache.poi.ss.usermodel.Row;
import org.apache.poi.ss.usermodel.Sheet;
import org.apache.poi.ss.usermodel.Workbook;
import org.apache.poi.xssf.usermodel.XSSFWorkbook;
public class PoiExcelRead {
 public static void main(String[] args) throws IOException {
     //文件名称
     File file = new File("data/c.xls");
     //File file = new File("data/c.xlsx");
     //根据文件名称获取操作工作簿
     Workbook workbook = getWorkBook(file);
     //获取读取的工作表，这里有两种方式
     Sheet sheet = workbook.getSheet("Sheet1");
     //Sheet sheet=workbook.getSheetAt(0);
     int allRow = sheet.getLastRowNum();//获取行数
     //按行读取数据
     for (int i = 0; i <= allRow; i++) {
        Row row = sheet.getRow(i);
        //获取列数
```

```java
        short lastCellNum = row.getLastCellNum();
        for (int j = 0; j < lastCellNum; j++) {
            String cellValue = row.getCell(j).getStringCellValue();
            System.out.print(cellValue + "\t");
        }
        System.out.println();
    }
    workbook.close();
}
/**
 * 判断Excel的版本，初始化不同的Workbook
 * @param in
 * @param filename
 * @return
 * @throws IOException
 */
public static Workbook getWorkBook(File file) throws IOException{
    //输入流
    InputStream in = new FileInputStream(file);
    Workbook workbook = null;
    //Excel 2003版本
    if(file.getName().endsWith("xls")){
        workbook = new HSSFWorkbook(in);
    // Excel 2007以上版本
    }else if(file.getName().endsWith("xlsx")){
        workbook = new XSSFWorkbook(in);
    }
    in.close();
    return workbook;
}
```

在程序 6-14 中提供了读取工作表的两种方法，分别是 getSheet(String name)方法和 getSheetAt(int index)方法，同时，程序中的 getLastRowNum()方法能够获取表格的所有行数，getLastCellNum()方法能够获取某行的所有列数。通过循环行数和列数，可以取出每个单元格的数据。图 6.17 所示为程序 6-14 在控制台的输出结果。

```
post_id  post_title
帖子id为: 00    帖子内容为: 0
帖子id为: 01    帖子内容为: 1
```

图 6.17　程序 6-14 的输出结果

6.2.3 爬虫案例

仍以采集 pythonforbeginners 网站中的帖子为例,介绍如何将网络爬虫获取的数据保存到 Excel 中。由于很多流程和 6.1.7 节相同,这里只进行简单介绍。首先,创建 PostModel,确定需要采集的字段,如程序 6-15 所示。

程序 6-15

```java
package com.crawler.pythonforbeginners;

public class PostModel {
    private String post_title;    //帖子标题
    private String post_content;  //帖子简介
    public String getPost_title() {
        return post_title;
    }
    public void setPost_title(String post_title) {
        this.post_title = post_title;
    }
    public String getPost_content() {
        return post_content;
    }
    public void setPost_content(String post_content) {
        this.post_content = post_content;
    }
}
```

接着,确定 URL 请求工具和网页内容解析工具,这里都使用 Jsoup。待数据解析完成后,利用 Jxl 或 POI 将所有数据插入 Excel,程序 6-16 给出了完整的代码。

程序 6-16

```java
package com.crawler.pythonforbeginners;
import java.io.IOException;
import java.util.ArrayList;
import java.util.List;
import org.jsoup.Jsoup;
import org.jsoup.nodes.Document;
import org.jsoup.nodes.Element;
import org.jsoup.select.Elements;
import jxl.write.WriteException;
import jxl.write.biff.RowsExceededException;
```

```java
public class CrawlerTest {

    public static void main(String[] args)
        throws IOException, RowsExceededException, WriteException {
        //待爬URL列表
        List<String> urlList = new ArrayList<String>();
        urlList.add("https://www.*****************.com/?page=1");
        urlList.add("https://www.*****************.com/?page=2");
        urlList.add("https://www.*****************.com/?page=3");
        List<PostModel> datalist = new ArrayList<PostModel>();
        //获取数据
        for (int i = 0; i < urlList.size(); i++) {
            datalist.addAll(crawerData(urlList.get(i)));
        }
        //存储数据
        DataToExcelByJxl.writeInfoListToExcel(
                "data/post.xls","sheet1",datalist);
        DataToExcelByPoi.writeInfoListToExcel(
                "data/post1.xlsx","sheet1",datalist);
    //解析数据
    static List<PostModel> crawerData(String url) throws IOException{
        //将所爬数据封装于集合中
        List<PostModel> datalist = new ArrayList<PostModel>();
        //获取URL对应的HTML内容
        Document doc = Jsoup.connect(url).timeout(30000).get();
        //定位需要采集的每个帖子
        Elements elements = doc.select("ul[class=nav nav-list]")
                .select("li[class=hentry]");
        //遍历每一个帖子
        for (Element ele : elements) {
            String post_title = ele.select(" h2 > a").text();
            String post_content = ele
                    .select("div[class=post-bodycopy cf]").text();
            //创建对象和封装数据
            PostModel model = new PostModel();
            model.setPost_title(post_title);
            model.setPost_content(post_content);
            datalist.add(model);
        }
        return datalist;
    }
}
```

在程序 6-16 中，调用了 DataToExcelByJxl 类中的 writeInfoListToExcel(String filePath,String sheetName,List<PostModel> datalist)方法将所有采集到的数据写入"post.xls"文件的"sheet1"表；调用了 DataToExcelByPoi 类中的 writeInfoListToExcel(String filePath,String sheetName,List<PostModel> datalist)方法将所有采集到的数据写入"post.xlsx"文件的"sheet1"表。程序 6-17 给出了 DataToExcelByJxl 类的内容，程序 6-18 给出了 DataToExcelByPoi 类的内容。

程序 6-17

```java
package com.crawler.pythonforbeginners;

import java.io.File;
import java.io.IOException;
import java.util.List;
import jxl.Workbook;
import jxl.write.Label;
import jxl.write.WritableSheet;
import jxl.write.WritableWorkbook;
import jxl.write.WriteException;
import jxl.write.biff.RowsExceededException;
public class DataToExcelByJxl {
 public static void writeInfoListToExcel(String filePath,
        String sheetName,List<PostModel> datalist)
    throws RowsExceededException, WriteException, IOException{
    File xlsFile = new File(filePath);
    // 创建一个工作簿
    WritableWorkbook workbook = Workbook.createWorkbook(xlsFile);
    // 创建一个工作表
    WritableSheet sheet = workbook.createSheet(sheetName, 0);
    //添加表头
    sheet.addCell(new Label(0, 0, "post_title"));
    sheet.addCell(new Label(1, 0, "post_content"));
    //添加内容
    for(int i=0;i<datalist.size();i++){
        sheet.addCell(new Label(0,i+1,
                datalist.get(i).getPost_title()));
        sheet.addCell(new Label(1,i+1,
                datalist.get(i).getPost_content()));
    }
    workbook.write();
    workbook.close();
```

```java
        System.out.println(">>>>>>>>>>数据写入完成!<<<<<<<<<<<<");
    }
}
```

程序6-18

```java
package com.crawler.pythonforbeginners;

import java.io.File;
import java.io.FileOutputStream;
import java.io.IOException;
import java.io.OutputStream;
import java.util.List;
import org.apache.poi.hssf.usermodel.HSSFWorkbook;
import org.apache.poi.ss.usermodel.Row;
import org.apache.poi.ss.usermodel.Sheet;
import org.apache.poi.ss.usermodel.Workbook;
import org.apache.poi.xssf.usermodel.XSSFWorkbook;
public class DataToExcelByPoi {
    public static void writeInfoListToExcel(String filePath,
            String sheetName,List<PostModel> datalist) throws IOException {
        //文件名称
        File file = new File(filePath);
        OutputStream outputStream = new FileOutputStream(file);
        Workbook workbook = getWorkBook(file);
        Sheet sheet = workbook.createSheet(sheetName);
        //添加表头
        Row row = sheet.createRow(0); //创建某行
        row.createCell(0).setCellValue("post_title");
        row.createCell(1).setCellValue("post_content");
        //添加内容
        for(int i = 0; i < datalist.size(); i++){
            Row everyRow = sheet.createRow(i + 1);
            everyRow.createCell(0).setCellValue(
                    datalist.get(i).getPost_title());
            everyRow.createCell(1).setCellValue(
                    datalist.get(i).getPost_content());
        }
        workbook.write(outputStream);
        //释放资源
        workbook.close();
```

```java
        outputStream.close();
        System.out.println(">>>>>>>>>>数据写入完成!<<<<<<<<<<<<");
    }
    /**
     * 判断Excel的版本,初始化不同的Workbook
     * @param filename
     * @return
     * @throws IOException
     */
    public static Workbook getWorkBook(File file) throws IOException{
        Workbook workbook = null;
        //Excel 2003版本
        if(file.getName().endsWith("xls")){
            workbook = new HSSFWorkbook();
            // Excel 2007以上版本
        }else if(file.getName().endsWith("xlsx")){
            workbook = new XSSFWorkbook();
        }
        return workbook;
    }
}
```

6.3 MySQL 数据存储

针对较大规模的数据,可使用数据库(Database)进行存储。数据库是基于数据结构来组织、存储和管理数据的仓库,每个数据库都有一个或多个不同的 API 用于创建、访问、管理、搜索和复制所存储的数据。目前,Oracle 旗下的 MySQL 是最流行的关系型数据库管理系统,其具有体积小、性能高、开源免费等特点。本节将介绍 MySQL 数据库相关的概念、基本用法、Java 操作数据库的方式(数据的读写、更新等操作);最后以具体的网络爬虫示例,演示如何将采集到的数据插入数据库。

在使用 MySQL 之前,需要下载和安装软件。本书中下载的版本为 MySQL Community Server 5.7。

在 Windows 系统中安装 MySQL 的过程都是通过界面化操作进行的,较为简单,在此不进行过多介绍。在 MySQL 安装完成后,可下载一款轻量级管理和设计数据库的工具 Navicat,其操作界面如图 6.18 所示。

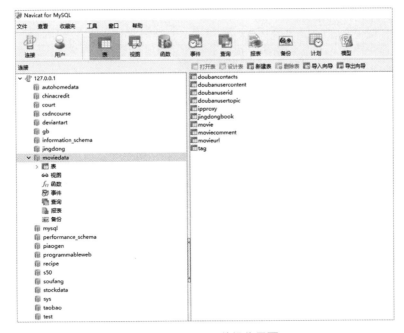

图 6.18　Navicat 的操作界面

6.3.1　数据库的基本概念

表 6.20 所示为数据库使用中的一些基本概念，包括数据库和数据表等。

表 6.20　数据库使用中的一些基本概念

术语	描述
数据库	用于存放多个数据表，将数据存储到数据表之前必须先建立数据库
数据表	存放于数据库中，类似于 Excel 表格
列	字段或属性，一列存放的是相同字段的内容，例如采集的评论字段
行	元素或记录，例如一条评论相关的内容（评论的 id、评论的时间、评论的内容等）
主键	数据表的唯一索引，例如评论的 id（能够唯一标识每一条记录）
索引	数据库某数据表中一个排序的数据结构，以协助快速查询、更新数据表中的数据
联合主键	数据表中的多个字段作为该数据表的主键

6.3.2　SQL 语句基础

SQL 的操作语句包含的内容较多，本节主要介绍一些基本操作。

1. 连接数据库

在 Windows 系统的 cmd 或者 Linux 环境中，连接数据库可使用如下命令。

```
mysql -u root -p
```

上述命令中的 root 表示用户，根据安装 MySQL 设置的用户名而定。输入此命令后，还需要输入密码，这样即可登录，如图 6.19 所示。

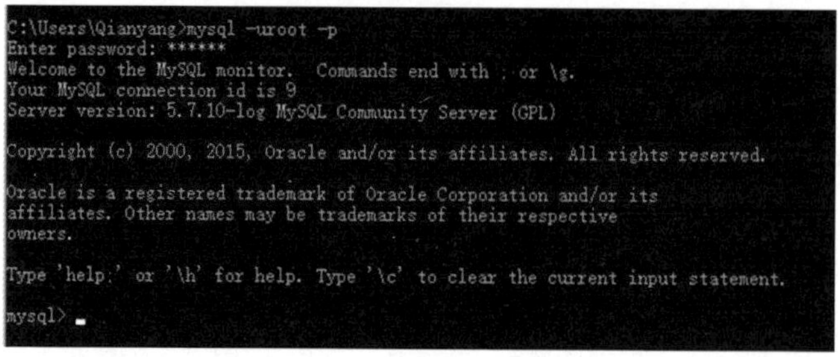

图 6.19　连接数据库命令

2. 查看已有数据库及使用某数据库

在 MySQL 中，可能包含多个已经创建好的数据库，可通过以下命令查看和使用已有的数据库。图 8.20 所示为查看和使用数据库命令。

```
show databases;
use test;
```

图 6.20　查看和使用数据库命令

3. 创建新的数据库及数据表

针对某特定的任务，如网络数据采集，需要创建指定的数据库存储一系列的数据。例如，通过如下命令创建一个 crawler 数据库。

```
create database crawler;
```

在数据库创建完成之后,可在数据库中进行建表。例如,我们想要存储网络爬虫采集的汽车销量数据,可通过以下命令在 crawler 数据库中创建 carsales 数据表。其中,COMMENT 用于解释说明某一字段,PRIMARY KEY 用于标识主键。图 6.21 所示为创建数据库及数据表命令。

```
use crawler;

CREATE TABLE IF NOT EXISTS `carsales` (
  `month` varchar(50) NOT NULL DEFAULT '' COMMENT '月份,例如2017-09-01',
  `sales` varchar(255) DEFAULT NULL COMMENT '该月份的汽车销量',
  PRIMARY KEY (`month`)
) ENGINE=InnoDB DEFAULT CHARSET=utf8;
```

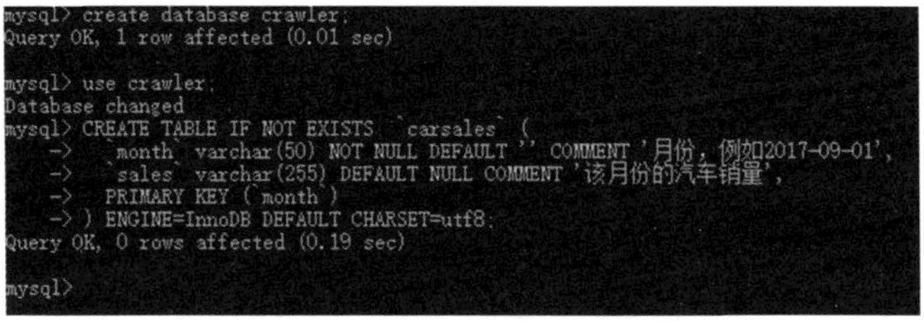

图 6.21 创建数据库及数据表命令

4. 查看数据库中的数据表及删除数据表

查看数据库中的数据表,使用的是 show 命令,删除某数据表使用的是 drop 命令,如图 6.22 所示。

```
show tables;

DROP TABLE  carsales;
#或者
DROP TABLE IF EXISTS carsales;
```

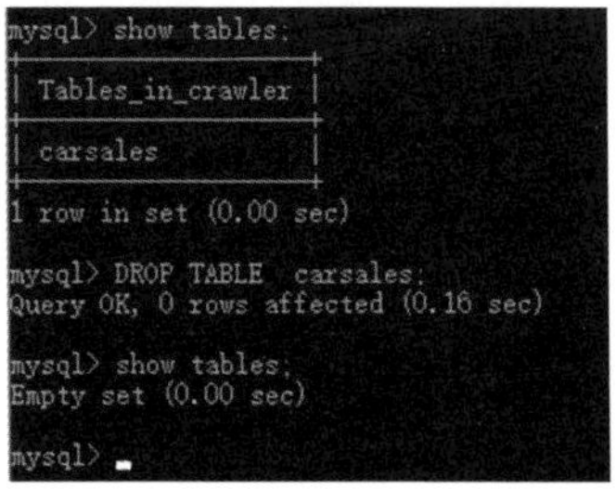

图 6.22 查看数据库中的数据表及删除数据表命令

5．向数据表中插入数据

向数据表中插入数据使用的是 insert 命令。

```
INSERT INTO carsales VALUES ('2017-09-01','500');
#或者
INSERT INTO carsales(month, sales) VALUES ('2017-09-01','500');
```

6．查看数据表中的数据

查询数据表中的数据使用的是 select 命令。图 6.23 所示为插入与查询命令。

```
#查询所有数据
select * from carsales;
#查询某列
select month from carsales;
```

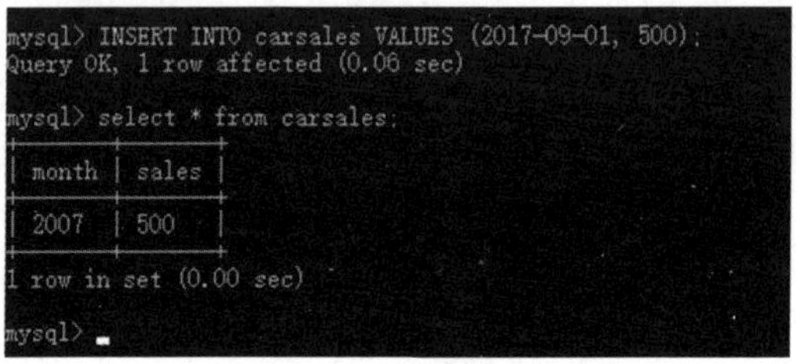

图 6.23　插入与查询命令

7．更新数据表中的数据

更新数据表中的数据使用的是 update 命令。图 6.24 所示为更新与查询命令。

```
update carsales set sales = '1000' where month = '2007';
```

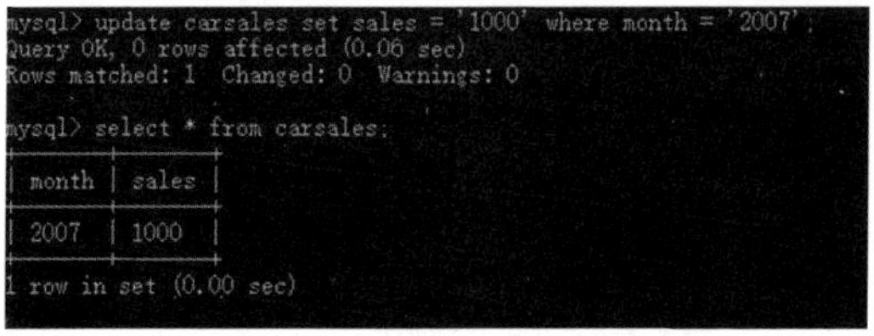

图 6.24　更新与查询命令

8．删除数据表中的数据

从数据表中删除数据使用的是 delete 语句。图 6.25 所示为删除与查询命令。

```
delete from carsales where month = '2007';
```

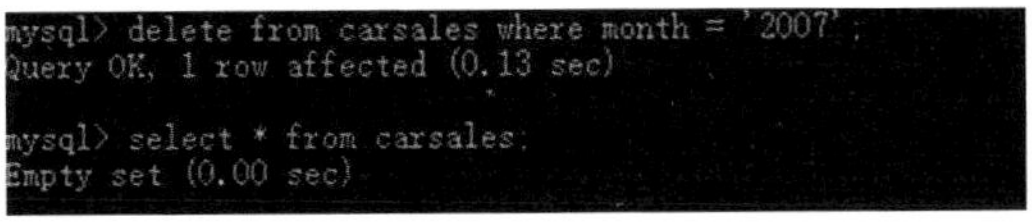

图 6.25　删除与查询命令

6.3.3 Java 操作数据库

1. 基本操作

Java 通过 JDBC（Java DataBase Connectivity）实现对数据库中的数据表的创建、查询和修改等一系列操作。如使用 Java 操作 MySQL 数据库需要下载相关的 JDBC 驱动程序 mysql-connector-java。利用 Maven 工程的 pom.xml 文件配置 jar 如下所示。

```xml
<!-- https:// *************.com/artifact/mysql/mysql-connector-java -->
<dependency>
    <groupId>mysql</groupId>
    <artifactId>mysql-connector-java</artifactId>
    <version>5.1.32</version>
</dependency>
```

Java 操作数据库的一般流程包括：

- 加载 JDBC 驱动；
- 基于数据库的地址、用户名以及密码，建立数据库连接；
- 创建 Statement 对象；
- 执行 SQL 语句；
- 关闭连接，释放资源。

程序 6-19 所示为 Java 按照上述流程操作数据库的简单示例。

程序 6-19

```java
import java.sql.Connection;
import java.sql.DriverManager;
import java.sql.ResultSet;
import java.sql.SQLException;
import java.sql.Statement;
public class MySQLConnections {
 private String driver = "";
 private String dbURL = "";
 private String user = "";
 private String password = "";
```

```java
    private static MySQLConnections connection = null;
    private MySQLConnections() throws Exception {
        driver = "com.mysql.jdbc.Driver";   //数据库驱动
        dbURL = "jdbc:mysql://127.0.0.1:3306/crawler"; //JDBC的URL
        user = "root";           //数据库用户名
        password = "112233";     //数据库密码
        System.out.println("dbURL:" + dbURL);
    }
    public static Connection getConnection() {
        Connection conn = null;
        if (connection == null) {
            try {
                connection = new MySQLConnections();  //初始化连接
            } catch (Exception e) {
                e.printStackTrace();
                return null;
            }
        }
        try {
            //调用Class.forName()方法加载驱动程序
            Class.forName(connection.driver);
            //建立数据库连接
            conn = DriverManager.getConnection(connection.dbURL,
                    connection.user, connection.password);
        } catch (Exception e) {
            e.printStackTrace();
        }
        return conn;
    }
    public static void main(String[] args) throws SQLException {
        Connection con = getConnection();
        Statement stmt = null;
        try {
            stmt = con.createStatement();//创建Statement对象
        } catch (SQLException e1) {
            e1.printStackTrace();
        }
        System.out.println("成功连接到数据库！");
        //防止数据库中有数据，先删除数据表中的数据
        stmt.execute("delete from carsales");
        //执行数据插入操作，忽略主键重复
        stmt.execute("insert ignore into carsales(month, sales)
```

```
values ('2017-09-01', '500')");
        stmt.execute("insert ignore into carsales(month, sales)
values ('2017-10-01', '100')");
        String sql = "select * from carsales";     //要执行的sql
        //查询操作
        ResultSet rs = stmt.executeQuery(sql);
        //结果集
        while (rs.next()){
            //输出1、2两列
            System.out.print(rs.getString(1) + "\t");
            System.out.print(rs.getString(2) + "\t");
            System.out.println();
        }
        stmt.addBatch("update carsales set sales = '1000' where month
= '2017-10-01' ");
        stmt.addBatch("update carsales set sales = '20' where month
= '2017-09-01'");
        int number[] = stmt.executeBatch();    //批处理
        System.out.println("执行的sql语句数目为:" + number.length);
        ResultSet rs1 = stmt.executeQuery("select * from carsales");
        System.out.println("更新后的结果为: ");
        while (rs1.next()){
            //输出1、2两列
            System.out.print(rs1.getString(1) + "\t");
            System.out.print(rs1.getString(2) + "\t");
            System.out.println();
        }
        con.close();    //关闭连接
    }
}
```

程序 6-19 中的 "127.0.0.1" 为数据库的 IP 地址，"3306" 为数据库的端口，crawler 为 Java 程序操作的数据库名。在与数据库建立连接后，需要通过 createStatement()方法实例化 Statement 对象，进而使用 Statement 类中的相关方法执行 SQL 语句，程序 6-20 列举了部分方法。程序 6-19 的输出结果如图 6.26 所示。

程序 6-20

```
ResultSet executeQuery(String sql) throws SQLException;   //获取结果集
int executeUpdate(String sql) throws SQLException;   //执行更新操作
//执行SQL语句,如更新
boolean execute(String sql) throws SQLException;
```

```
//用于批处理，和executeBatch()方法前后使用
void addBatch( String sql ) throws SQLException;
int[] executeBatch() throws SQLException;                    //批处理
```

```
dbURL:jdbc:mysql://127.0.0.1:3306/crawler
成功连接到数据库！
2017-09-01      500
2017-10-01      100
执行的sql语句数目为:2
更新后的而结果为：
2017-09-01      20
2017-10-01      1000
```

图 6.26　程序 6-19 的输出结果

2. 基于 XML 配置文件连接数据库

在实际应用中，经常需要配置多个数据库，以实现跨库数据操作。例如，使用 Spring 和 Mybatis 框架构建网站系统时，需要通过配置文件 jdbc.propertis 实现多数据源配置。同样，针对网络爬虫系统，也可利用配置文件配置多个数据库。下面，将使用 XML 配置文件实现数据库的配置。首先，在工程的根目录下创建 db.xml 文件（见图 6.27），并在该文件中添加 JDBC 相关的参数，如程序 6-21 所示。

程序 6-21

```xml
<?xml version="1.0" encoding="UTF-8"?>
<config>
 <connectionInfo>
    <node1>
        <nodeName>node1</nodeName>
        <driver-name>com.mysql.jdbc.Driver</driver-name>
        <url>jdbc:mysql://127.0.0.1:3306/crawler</url>
        <username>root</username>
        <password>112233</password>
    </node1>

    <node2>
        <nodeName>node2</nodeName>
        <driver-name>com.mysql.jdbc.Driver</driver-name>
        <url>jdbc:mysql://114.213.252.26:3306/crawler</url>
        <username>root</username>
        <password>112233</password>
    </node2>
 </connectionInfo>
</config>
```

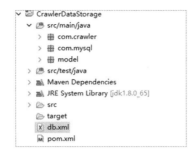

图 6.27　创建 db.xml 文件

在程序 6-21 的配置文件中，共添加了两个数据库节点，其数据库地址分别是"127.0.0.1"和"114.213.252.26"。为了读取和解析 XML 配置文件的内容，我们需要下载 Dom4j（依赖 jaxen），下面利用 Maven 工程的 pom.xml 文件配置相应的 jar 包。

```xml
<!-- https://************.com/artifact/dom4j/dom4j -->
<dependency>
    <groupId>dom4j</groupId>
    <artifactId>dom4j</artifactId>
    <version>1.6.1</version>
</dependency>
<!-- https://************.com/artifact/jaxen/jaxen -->
<dependency>
    <groupId>jaxen</groupId>
    <artifactId>jaxen</artifactId>
    <version>1.1.6</version>
</dependency>
```

程序 6-22 所示为利用 XML 配置文件连接数据库的操作案例，即读取数据库配置文件，解析数据库的基本参数，最后建立数据库连接。图 6.28 所示为程序 6-22 的输出结果。

程序 6-22

```java
import java.io.File;
import java.sql.Connection;
import java.sql.DriverManager;
import java.sql.ResultSet;
import java.sql.SQLException;
import java.sql.Statement;
import org.dom4j.Document;
import org.dom4j.Element;
import org.dom4j.io.SAXReader;
```

```java
public class XMLMySQLConnections {

    private String driver = "";
    private String dbURL = "";
    private String user = "";
    private String password = "";
    private static XMLMySQLConnections connection = null;
    private XMLMySQLConnections(String node) throws Exception {
        //读取XML文件
        SAXReader reader=new SAXReader();
        Document doc=reader.read(new File("db.xml"));
        //取得JDBC相关的配置
        Element driverNameEle = (Element)doc.selectObject("/config/connectionInfo/" + node + "/driver-name");
        Element urlEle = (Element) doc.selectObject("/config/connectionInfo/" + node + "/url");
        Element usenameEle = (Element)doc.selectObject("/config/connectionInfo/" + node + "/username");
        Element passwordEle = (Element) doc.selectObject("/config/connectionInfo/" + node + "/password");
        driver = driverNameEle.getStringValue();         //数据库驱动
        dbURL = urlEle.getStringValue();                  //JDBC的URL
        user = usenameEle.getStringValue();               //数据库用户名
        password = passwordEle.getStringValue();          //数据库密码
    }
    public static Connection getConnection(String node) {
        Connection conn = null;
        if (connection == null) {
            try {
                connection = new XMLMySQLConnections(node);   //初始化
            } catch (Exception e) {
                e.printStackTrace();
                return null;
            }
        }
        try {
            //调用Class.forName()方法加载驱动程序
            Class.forName(connection.driver);
            //建立数据库连接
            conn = DriverManager.getConnection(connection.dbURL,
                    connection.user, connection.password);
        } catch (Exception e) {
```

```java
            e.printStackTrace();
        }
        return conn;
    }
    public static void main(String[] args) throws SQLException {
        Connection con = getConnection("node1");//需要连接的节点数据库
        Statement stmt = con.createStatement();//创建Statement对象
        System.out.println("成功连接到数据库! ");
        String sql = "select * from carsales"; //要执行的SQL
        ResultSet rs = stmt.executeQuery(sql); //结果集
        while (rs.next()){
            //输出1、2两列
            System.out.print(rs.getString(1) + "\t");
            System.out.print(rs.getString(2) + "\t");
            System.out.println();
        }
        con.close();   //关闭连接
    }
}
```

```
成功连接到数据库!
2017-09-01        20
2017-10-01        1000
```

图 6.28　程序 6-22 的输出结果

3. 基于 QueryRunner 操作数据库

下面，介绍一个非常好用的数据库操作类 QueryRunner（来源于 dbutils 工具包），该类使得 JDBC 编程更加方便与快捷，并且线程操作也是安全的。在使用 QueryRunner 前，先使用 pom.xml 文件配置相关的 jar 包。

```xml
<!-- QueryRunner 相关依赖   -->
<dependency>
    <groupId>commons-dbutils</groupId>
    <artifactId>commons-dbutils</artifactId>
    <version>1.7</version>
</dependency>
<dependency>
    <groupId>org.apache.commons</groupId>
    <artifactId>commons-dbcp2</artifactId>
    <version>2.5.0</version>
</dependency>
```

QueryRunner 类中涉及一系列操作数据库的方法，以下列举三种最为常用的方法，即更新、查询和批处理。

```
//执行insert、update、delete对应的SQL语句
public int update(String sql) throws SQLException
//泛型方法：执行SELECT查询语句。如查询数据表中的某列数据时，可获得该列集合
public <T> T query(String sql, ResultSetHandler<T> rsh) throws SQLException
//泛型方法：执行insert、update、delete批处理操作
public int[] batch(String sql, Object[][] params) throws SQLException
```

QueryRunner 在连接数据库时，使用 dbcp2 包中的 BasicDataSource，该类的作用是配置数据库的基本信息，如驱动、用户名和密码等。基于 6.3.2 节中的汽车销量表 carsales，构建 CarSaleModel 类，如程序 6-23 所示。构建 CarSaleModel 类的作用是为了实现泛型查询操作。程序 6-24 演示了 update()和 query()方法的使用情况，具体包括更新 carsales 数据表的某行数据，查询某行数据以及查询多行数据。图 6.29 所示为程序 6-24 的输出结果。

程序 6-23

```java
public class CarSaleModel {
    private String month;
    private String sales;
    public String getMonth() {
        return month;
    }
    public void setMonth(String month) {
        this.month = month;
    }
    public String getSales() {
        return sales;
    }
    public void setSales(String sales) {
        this.sales = sales;
    }
}
```

程序 6-24

```java
import java.sql.SQLException;
import java.util.List;
import javax.sql.DataSource;
```

```java
import org.apache.commons.dbcp2.BasicDataSource;
import org.apache.commons.dbutils.QueryRunner;
import org.apache.commons.dbutils.handlers.BeanListHandler;
import org.apache.commons.dbutils.handlers.ColumnListHandler;
import model.CarSaleModel;
public class QueryRunnerTest {
    //获取数据库信息
    static DataSource ds = getDataSource("jdbc:mysql://127.0.0.1:3306/crawler");
    //使用QueryRunner类操作数据库
    static QueryRunner qr = new QueryRunner(ds);
    public static DataSource getDataSource(String connectURI){
        BasicDataSource ds = new BasicDataSource();
        //MySQL的JDBC驱动
        ds.setDriverClassName("com.mysql.jdbc.Driver");
        ds.setUsername("root");          //所要连接的数据库名
        ds.setPassword("112233");         //MySQL的登录密码
        ds.setUrl(connectURI);
        return ds;
    }
    // update方法
    public void executeUpdate(String sql){
        try {
            qr.update(sql);
        } catch (SQLException e) {
            e.printStackTrace();
        }
    }
    //按照SQL查询多个结果,这里需要构建Bean对象
    public <T> List<T> getListInfoBySQL (String sql, Class<T> type ){
        List<T> list = null;
        try {
            list = qr.query(sql,new BeanListHandler<T>(type));
        } catch (SQLException e) {
            e.printStackTrace();
        }
        return list;
    }
    //查询一列
    @SuppressWarnings({ "unchecked", "rawtypes" })
    public List<Object> getListOneBySQL (String sql,String id){
        List<Object> list=null;
```

```java
        try {
            list = (List<Object>) qr.query(sql, new ColumnListHandler(id));
        } catch (SQLException e) {
            e.printStackTrace();
        }
        return list;
    }
    public static void main(String[] args) {
        QueryRunnerTest QueryRunnerTest = new QueryRunnerTest();
        //查询多列，这里利用了CarSaleModel类
        List<CarSaleModel> mutllistdata = QueryRunnerTest.getListInfoBySQL("select month,"
                + "sales from carsales", CarSaleModel.class);
        System.out.println("更新前的数据为:");
        for (CarSaleModel model : mutllistdata) {
            System.out.println(model.getMonth() + "\t" +model.getSales());
        }
        //执行更新操作
        QueryRunnerTest.executeUpdate("update carsales set sales = '4000' "
                + "where month = '2017-10-01'"); //执行更新操作
        System.out.println("更新操作完成!");
        //查询多列
        List<CarSaleModel> mutllistdataupdate = QueryRunnerTest.getListInfoBySQL("select "
                + "month,sales from carsales", CarSaleModel.class);
        for (CarSaleModel model : mutllistdataupdate) {
            System.out.println(model.getMonth() + "\t" +model.getSales());
        }
        System.out.println("读取多列数据完成");
        //查询单列
        List<Object> listdata = QueryRunnerTest.getListOneBySQL("select sales from "
                + "carsales", "sales");
        for (int i = 0; i < listdata.size(); i++) {
            System.out.println(listdata.get(i).toString());
        }
```

```
            System.out.println("按列读取数据完成! ");
        }
    }
```

```
更新前的数据为:
2017-09-01        20
2017-10-01        1000
更新操作完成!
2017-09-01        20
2017-10-01        4000
读取多列数据完成
20
4000
按列读取数据完成!
```

图 6.29　程序 6-24 的输出结果

6.3.4　爬虫案例

本节仍以某汽车网站上某汽车销量采集为例，演示数据采集入库。详细流程包括：

- 确定需要采集的信息，定义具体的 JavaBean。如本案例采集的字段包括时间与销量，因此定义了 CarSaleModel 类（见程序 6-22）。
- 确定需要采集的 URL（或 URL 列表），使用 Jsoup 获取网页的内容并解析具体的字段。
- 将解析后的数据封装到集合中。
- 将封装后的数据插入指定的数据表。此步骤涉及数据库的连接，可直接利用程序 6-24 进行操作。
- 为执行数据批量插入操作，可在 QueryRunnerTest 类中添加一种新方法，如程序 6-25 所示。

程序 6-25

```
//将所爬数据插入数据库
public static  void insertData ( List<CarSaleModel> datalist ) {
    Object[][] params = new Object[datalist.size()][2];
    for ( int i = 0; i < datalist.size(); i++ ){
        //需要存储的字段
        params[i][0] = datalist.get(i).getMonth();
        params[i][1] = datalist.get(i).getSales();
    }
    QueryRunner qr = new QueryRunner(ds);
    try {
        //执行批处理语句
        qr.batch("INSERT INTO carsales(month,sales) VALUES (?,?)", params);
```

```
    } catch (SQLException e) {
        System.out.println(e);
    }
    System.out.println("新闻数据入库完毕");
}
```

该方法的输入是 List<CarSaleModel>集合，使用的是 QueryRunner 类中的 batch() 方法。执行程序 6-26 数据采集入库的主程序，可将该车型每月的销量插入数据库，如图 6.30 所示。

程序 6-26

```java
import java.io.IOException;
import java.util.ArrayList;
import java.util.List;
import org.jsoup.Jsoup;
import org.jsoup.nodes.Document;
import org.jsoup.nodes.Element;
import org.jsoup.select.Elements;
import com.mysql.QueryRunnerTest;
import model.CarSaleModel;
public class CrawlerToDatabaseTest {
    public static void main(String[] args) throws IOException {
        //获取URL对应的XML内容
        Document doc = Jsoup.connect("http://db.auto.****.com/cxdata/xml/sales/model/model1001sales.xml")
                .timeout(5000).get();
        //数据存储到集合中
        List<CarSaleModel> datalist = new ArrayList<CarSaleModel>();
        //Jsoup选择器解析
        Elements sales_ele = doc.select("sales");
        for (Element elem:sales_ele) {
            String salesnum=elem.attr("salesnum");
            String date = elem.attr("date");
            //封装对象
            CarSaleModel model = new CarSaleModel();
            model.setMonth(date);
            model.setSales(salesnum);
            //添加到集合中
            datalist.add(model);
            System.out.println("月份:" + date + "\t销量:" + salesnum);
```

```
        }
        //将所爬数据插入数据库
        QueryRunnerTest.insertData(datalist);
    }
}
```

month	sales
2007-01-01	14834
2007-02-01	9687
2007-03-01	18173
2007-04-01	18508
2007-05-01	19710
2007-06-01	20311
2007-07-01	17516
2007-08-01	17535
2007-09-01	17743
2007-10-01	15255
2007-11-01	17250
2007-12-01	14609

图 6.30 汽车销量采集

6.4 本章小结

本章介绍了网络爬虫数据存储的几种方式，包括输入流与输出流、Excel 和数据库的使用。针对图片、PDF 和音频等文件的存储，通常使用的是输入流与输出流存储；针对小规模数据存储，可使用 Excel 存储；针对较大规模数据存储，需要使用数据库软件存储，常用的是 MySQL。

第 7 章

网络爬虫实战项目

本章将介绍一个简单的网络爬虫框架。通过对该框架的学习,读者可以掌握 Java 网络爬虫项目编写的流程。以下将通过三个具体的网络爬虫项目展示此框架的使用情况。

7.1 新闻数据采集

7.1.1 采集的网页

本项目将以国外某财经新闻网站的新闻采集为例,展示网络爬虫框架的使用情况。

网址对应的内容如图 7.1 所示。将网页内容下拉到最底部,会发现翻页按钮,如图 7.2 所示。

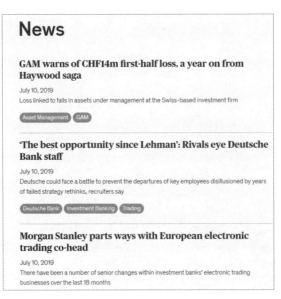

图 7.1 国外某财经新闻网站新闻展示

图 7.2 翻页按钮

在编写网络爬虫之前,需要通过网络抓包分析待采集的页面。使用谷歌浏览器对第一页进行抓包分析,结果如图 7.3 所示。从图 7.3 中可以看出该网页中的新闻信息存放在 JSON 字符串中,并且该 JSON 字符串通过<script>元素嵌入 HTML。

图 7.3 网络抓包分析结果

将该 JSON 字符串复制到在线校准网站中,可以发现字符串中的 collection 字段包含了所有的新闻 id,如图 7.4 所示。

图 7.5 所示为 JSON 字符串中的每条新闻的存储方式。从图 7.5 中可以发现所要解析的相关新闻的字段(如新闻标题、新闻标签、新闻摘要等),需要先解析得到 collection 字段中的 id(如 FN5000054754),利用 id 拼接形如"article_FN5000054754"的字符串。

另外,在实现网络爬虫的翻页操作时,只需要改变 URL 中最后一个字符即可,如第 2 页数据对应的 URL 为 https://www.********.com/news/2。

图 7.4　collection 字段中的新闻 id

图 7.5　JSON 字符串中的新闻的存储方式

7.1.2　框架介绍

基于网络爬虫原理和开发逻辑，笔者设计了一个框架，利用此框架可以轻松地开发一些简单的网络爬虫。同时，读者也可在此框架基础上继续添加相关的 package 和类，开发一些较为复杂的网络爬虫。

图 7.6 所示为框架的结构，其包含 5 个 package，即 com.(db、main、model、parse、util)，每个 package 都有其独特的作用。

com.db 主要存放的是数据库操作类，包含 MyDataSource.java 和 MYSQLControl.java 两个文件。MyDataSource.java 负责数据库驱动注册、设置连接数据库的用户名和密码，MYSQLControl.java 负责数据库连接、数据表构建、数据插入和更新等。

com.model 存放的是相关实体类，用来封装数据。例如，采集新闻的标题、简介、URL、标签等内容，则需要在对应的 MODEL 类中写入相应的属性。

com.util 主要存放 URL 请求类、时间操作类和文本读写类等，如 HttpRequestUtil 和 TimeUtil，其功能是获取指定 URL 对应的 HTML、XML 和 JSON 内容；对相关字段进行处理（如时间标准化）。

com.parse 主要负责对 com.util 中获取的 HTML、XML 和 JSON 等内容进行**解析**。如 HTML 及 XML 可以采用 Jsoup 解析、JSON 数据可采用 Fastjson 解析。

com.main 存放主程序。包括调用 com.util 中的类与方法获取网页内容，调用 com.parse 中的类与方法解析得到具体的数据，调用 com.db 中的类与方法将采集到的数据存储到数据库。

图 7.6　框架的结构

7.1.3　程序编写

下面将利用上述框架，实现新闻数据的采集。其程序编写流程如下所示。

1. jar 包的下载

基于 Maven 工程中的 pom.xml 下载该项目需要的相关 jar 包。

```xml
<!--- httpclient 相关依赖 jar 包 -->
<dependency>
    <groupId>org.apache.httpcomponents</groupId>
    <artifactId>httpclient</artifactId>
    <version>4.5.5</version>
</dependency>
<!-- 数据库相关 -->
<dependency>
    <groupId>mysql</groupId>
    <artifactId>mysql-connector-java</artifactId>
    <version>5.1.32</version>
</dependency>
<dependency>
    <groupId>commons-dbutils</groupId>
    <artifactId>commons-dbutils</artifactId>
    <version>1.7</version>
</dependency>
<dependency>
    <groupId>org.apache.commons</groupId>
    <artifactId>commons-dbcp2</artifactId>
    <version>2.5.0</version>
</dependency>
<!-- Json 解析 Fastjson -->
<dependency>
    <groupId>com.alibaba</groupId>
    <artifactId>fastjson</artifactId>
    <version>1.2.47</version>
</dependency>
```

2. com.model

明确要采集的字段，编写相应的实体类。例如，本项目采集的数据字段包括新闻 id、新闻 URL、新闻标题、新闻时间、新闻简介和新闻标签。因此，可构建 NewsModel 类，如程序 7-1 所示。

另外，还需要创建 CollocationModel，用于封装如图 7.4 所示的 collection 字段中的新闻 id，如程序 7-2 所示。

程序 7-1

```java
package com.model;
//需要采集的数据字段，该字段要与JSON中的字段名称保持一致
public class NewsModel {
    private String id;              //新闻id
    private String headline;        //新闻标题
    private String url;             //新闻URL
    private String timestamp;       //新闻时间
    private String tags;            //新闻标签
    private String summary;         //新闻简介
    public String getId() {
        return id;
    }
    public void setId(String id) {
        this.id = id;
    }
    public String getHeadline() {
        return headline;
    }
    public void setHeadline(String headline) {
        this.headline = headline;
    }
    public String getUrl() {
        return url;
    }
    public void setUrl(String url) {
        this.url = url;
    }
    public String getTimestamp() {
        return timestamp;
    }
    public void setTimestamp(String timestamp) {
        this.timestamp = timestamp;
    }
    public String getTags() {
        return tags;
    }
    public void setTags(String tags) {
        this.tags = tags;
    }
    public String getSummary() {
```

```java
        return summary;
    }
    public void setSummary(String summary) {
        this.summary = summary;
    }

}
```

程序 7-2

```java
package com.model;

public class CollocationModel {
    private String id;

    public String getId() {
        return id;
    }

    public void setId(String id) {
        this.id = id;
    }
}
```

3. com.util

com.util 中存放有 URL 请求类 HttpRequestUtil。给定 URL，利用该类便可以获得相应的 HTML、XML 和 JSON 内容，该类使用 HttpClient 工具编写，其代码如程序 7-3 所示。

程序 7-3

```java
package com.util;
import java.io.IOException;
import java.util.ArrayList;
import java.util.List;
import org.apache.http.Header;
import org.apache.http.HttpEntity;
import org.apache.http.HttpHeaders;
import org.apache.http.HttpResponse;
import org.apache.http.client.ClientProtocolException;
import org.apache.http.client.HttpClient;
import org.apache.http.client.methods.HttpGet;
```

```java
import org.apache.http.impl.client.DefaultHttpRequestRetryHandler;
import org.apache.http.impl.client.HttpClients;
import org.apache.http.message.BasicHeader;
import org.apache.http.util.EntityUtils;
public class HttpRequestUtil {
    private HttpClient httpClient;
    private List<Header> headerList = new ArrayList<Header>();
    public HttpEntity getEntityByHttpGetMethod(String url){
        initDefaultHeaders(); //头信息
        httpClient = HttpClients.custom()
                //默认重试次数
                .setRetryHandler(new DefaultHttpRequestRetryHandler())
                .setDefaultHeaders(headerList)   //添加头信息
                .build();
        //请求中的cookie策略,如果不添加,则容易出现Invalid cookie header警告
        RequestConfig defaultConfig = RequestConfig.custom()
                .setCookieSpec(CookieSpecs.STANDARD).build();
        HttpGet httpget = new HttpGet(url);      //使用的请求方法
        httpget.setConfig(defaultConfig);
        HttpResponse response = null;
        try {
            response = httpClient.execute(httpget);
        } catch (ClientProtocolException e) {
            e.printStackTrace();
        } catch (IOException e) {
            e.printStackTrace();
        }
        HttpEntity httpEntity = response.getEntity(); //获取网页内容流
        return httpEntity;
    }
    //指定URL及网页编码,获取网页内容
    public String getContentByHttpGetMethod(String url,String code){
        try {
            return EntityUtils.toString(getEntityByHttpGetMethod(url),code);
        } catch (IOException e) {
            e.printStackTrace();
            return null;
        }
    }
    //头信息设置
```

```java
    private List<Header> initDefaultHeaders(){
        headerList.add(new BasicHeader(HttpHeaders.ACCEPT,
                "text/html,application/xhtml+xml,application/xml;q=0.9," +
                "image/webp,image/apng,*/*;q=0.8"));
        headerList.add(new BasicHeader(HttpHeaders.USER_AGENT,
                "Mozilla/5.0 (Windows NT 10.0; Win64; x64) " +
                "AppleWebKit/537.36 (KHTML, like Gecko)"
                + " Chrome/60.0.3112.113 Safari/537.36"));
        headerList.add(new BasicHeader(HttpHeaders.ACCEPT_ENCODING,
                "gzip, deflate"));
        headerList.add(new BasicHeader(HttpHeaders.CACHE_CONTROL,
                "max-age=0"));
        headerList.add(new BasicHeader(HttpHeaders.CONNECTION,
                "keep-alive"));
        headerList.add(new BasicHeader(HttpHeaders.ACCEPT_LANGUAGE,
                "zh-CN,zh;q=0.8,en;q=0.6,zh-TW;q=0.4,ja;q=0.2," +
                "de;q=0.2"));
        return headerList;
    }
}
```

HttpRequestUtil 的使用具有一般性，对大多数 URL 的请求皆适用，如程序 7-4 所示。首先，实例化 HttpRequestUtil，之后调用 getContentByHttpGetMethod()方法执行 URL 请求，其中，"gbk"指的是网页的编码。

程序 7-4

```java
HttpRequestUtil httpRequest = new HttpRequestUtil();
String url = "https://www.********.com/news/1";
//调用HttpRequest中的方法获取网页内容
String htmlString= httpRequest.getContentByHttpGetMethod(url,"gbk");
```

另外，在该网页对应的 JSON 字符串中，字段 timestamp 使用的是 UNIX 时间戳。为便于数据的存储及查询，这里需要编写时间操作类 TimeUtil，该类中的 TimeStampToDate()方法用于将 UNIX 时间戳转化为指定格式的时间（已在 2.8 节中介绍）。TimeUtil 类的代码如程序 7-5 所示。

程序 7-5

```java
package com.util;
import java.text.SimpleDateFormat;
```

```
import java.util.Date;
import java.util.Locale;
public class TimeUtil {
  public static String TimeStampToDate(String timestampString, String formats) {
        //注意这里使用的是毫秒
        Long timestamp = Long.parseLong(timestampString);
        String date = new SimpleDateFormat(formats, Locale.CHINA).format(new Date(timestamp));
        return date;
  }
}
```

4. com.parse

在使用 HttpRequestUtil 类请求 URL 时，获得的响应实体为 HTML 类型的字符串。为提取<script>标签中的 JSON 字符串，这里使用了 String 类中的 split()和 substring() 方法。针对已标准化的 JSON 字符串，再使用 Fastjson 工具进行解析。

基于对 JSON 字符串的分析，编写解析 Parse 类，如程序 7-6 所示。在程序 7-6 中，首先使用了 Fastjson 中 parseObject(String text, Feature... features)方法，将 JSON 字符串转化成 JSON 对象（注意：该 JSON 字符串为复杂嵌套式 JSON，为防止转化过程的 JSON 乱序导致解析失败，才使用此种方法）。之后，调用 Fastjson 中操作 JSON 数组的 parseArray(String text, Class<T> clazz)方法解析 collection 字段中的新闻 id 集合。最后，遍历每个新闻 id，调用 Fastjson 中操作 JSON 对象的 parseObject(String text, Class<T> clazz)方法解析每条新闻的字段。

程序 7-6

```
package com.parse;
import java.util.ArrayList;
import java.util.List;
import com.alibaba.fastjson.JSON;
import com.alibaba.fastjson.JSONObject;
import com.model.CollocationModel;
import com.model.NewsModel;
public class Parse {
    //解析JSON数据
    public static List<NewsModel> getData (String html, int page) {
        //创建集合，用于封装每个页面的所有新闻数据
        List<NewsModel> dataList = new ArrayList<NewsModel>();
        //获取<script></script>中的JSON字符串
```

```java
        String getScriptJson = html.split(" window.__STATE__ = ")[1]
                .split("</script>")[0].trim();
    //转化成标准JSON文件
    String standardJson = getScriptJson
            .substring(0, getScriptJson.length() - 1);
     //注意这里在解析时,不能调整顺序,否则可能会报错
    JSONObject jsonfile = JSONObject.parseObject(standardJson,
Feature.OrderedField);
    //拼接search_collection,用于获取所有新闻id
    String search_collection = "search_collection_"
            + "{\"db\":\"fnonline\",\"query"
            + "\":\"news\",\"queryType\""
            + ":\"keyword\",\"page\":" + page + "}" ;

    //解析collection字段中的所有的新闻id,这里需要调用parseArray()方法
    List<CollocationModel> idCollection = parseArray(
            JSONObject.parseObject(JSONObject.parseObject(
            JSONObject.parseObject(jsonfile.get("data")
                    .toString()).get(search_collection)
                    .toString()).get("data")
                    .toString()).get("collection")
                    .toString(), CollocationModel.class);
    for (CollocationModel id : idCollection) {
        //拼接形如article_FN5000054754的字符串
        String article = "article_" + id.getId();
        //获取每一条新闻对应的JSON数据
        String oneNewsString = JSONObject.parseObject(
                JSONObject.parseObject(
                JSONObject.parseObject(jsonfile.get("data")
                        .toString()).get(article)
                        .toString()).get("data")
                        .toString()).get("data")
                        .toString();
        NewsModel   model   =   JSON.parseObject(oneNewsString,
NewsModel.class);
        dataList.add(model);
    }
    //返回集合
    return dataList;
}

/**
```

```
 * 解析JsonArray数据
 *
 * @param jsonString
 * @param cls
 * @return
 */
public static <T> List<T> parseArray(String jsonString,
Class<T> cls) {
    List<T> list = new ArrayList<T>();
    try {
        list = JSON.parseArray(jsonString, cls);
    } catch (Exception e) {
        e.printStackTrace();
    }
    return list;
}
}
```

5. com.db

com.db 中包含两个文件，即 MyDataSource 和 MYSQLControl。MyDataSource 用来配置数据库的基本信息；MYSQLControl 负责操作数据库。为了编写方便，这里直接使用数据库操作类 QueryRunner。

在此步骤中，需要将 com.parse 解析后获得的集合数据，插入指定数据库的数据表。在本节中，使用的数据库名为 crawler，创建的数据表为 fnlondonnews，其建表语句如下所示。

```
CREATE TABLE `fnlondonnews` (
  `id` varchar(50) NOT NULL,
  `headline` varchar(200) DEFAULT NULL,
  `url` varchar(200) DEFAULT NULL,
  `time` datetime DEFAULT NULL,
  `tags` varchar(200) DEFAULT NULL,
  `summary` text,
  PRIMARY KEY (`id`)
) ENGINE=InnoDB DEFAULT CHARSET=utf8;
```

程序 7-7 为 MyDataSource 类的编写，即添加了数据库的用户名和密码，并使用 setUrl() 方法接收数据库连接地址。

程序 7-7

```java
package com.db;

import javax.sql.DataSource;
import org.apache.commons.dbcp2.BasicDataSource;
public class MyDataSource {
  public static DataSource getDataSource(String connectURI){
      BasicDataSource ds = new BasicDataSource();
      //MySQL的JDBC驱动
      ds.setDriverClassName("com.mysql.jdbc.Driver");    //驱动名称
      ds.setUsername("root");                            //数据库用户名
      ds.setPassword("112233");                          //数据库密码
      ds.setUrl(connectURI);                             //连接地址
      return ds;
  }
}
```

程序 7-8 为 MYSQLControl 类的编写。这里首先调用了 MyDataSource 类中的 getDataSource()方法来创建数据库连接。为将采集的集合数据，插入到数据库，这里又编写了 executeInsert(List<NewsModel> data)方法，该方法使用了 QueryRunner 类中的 batch()方法来实现数据的批量插入。

程序 7-8

```java
package com.db;

import java.sql.SQLException;
import java.util.List;
import javax.sql.DataSource;
import org.apache.commons.dbutils.QueryRunner;
import com.model.NewsModel;
import com.util.TimeUtil;
public class MYSQLControl {
   //根据本地的数据库地址修改
    static DataSource ds = MyDataSource.getDataSource("jdbc:mysql://127.0.0.1:3306/"
            + "crawler?useUnicode=true&characterEncoding=UTF8");
    static QueryRunner qr = new QueryRunner(ds);
    public static void executeInsert(List<NewsModel> data)  {
```

```
            Object[][] params = new Object[data.size()][6];  //数据的维度
            for ( int i = 0; i < params.length; i++ ){
                params[i][0] = data.get(i).getId();
                params[i][1] = data.get(i).getHeadline();
                params[i][2] = data.get(i).getUrl();
                params[i][3] = TimeUtil.TimeStampToDate(
                        data.get(i).getTimestamp(),
                        "yyyy-MM-dd HH:mm:ss");  //时间标准化
                params[i][4] = data.get(i).getTags();
                params[i][5] = data.get(i).getSummary();
            }
            //使用batch方法批量插入
            try {
                qr.batch("insert into  fnlondonnews(id,headline,url,"
                        + "time,tags,summary)"
                        + "values (?,?,?,?,?,?)", params);
                System.out.println("执行数据库完毕！"+"成功插入数据：
"+data.size()+"条");
            } catch (SQLException e) {
                e.printStackTrace();
            }
        }
    }
}
```

6．com.main

com.main 包用于存放工程的主类。在本案例中，需要采集的网页包含多页，所以可以通过拼接 URL 的方式获取每页内容。

```
String  everypageurl = "https://www.********.com/news/" + page ;
```

在上述代码中，page 为所要采集的页面编号，如 1、2……。在编写具有翻页功能的网络爬虫时，经常使用这种方式拼接每页的 URL。程序 7-9 所示为主类 CrawlerMain。

程序 7-9

```
package com.main;

import java.util.List;
import com.db.MYSQLControl;
import com.model.NewsModel;
```

```java
import com.parse.Parse;
import com.util.HttpRequestUtil;
public class CrawlerMain {
  static HttpRequestUtil httpRequest = new HttpRequestUtil();
  public static void main(String[] args) throws Exception {
      //调用HttpRequest中的方法获取网页内容
      for(int i = 1; i < 5; i++){
          int page = i;  //爬取的页数
          //拼接URL,实现翻页操作
          String everypageurl = "https://www.********.com/news/" + page ;
          //调用HttpRequest中的方法获取网页内容
          String html = httpRequest.getContentByHttpGetMethod(everypageurl,"gbk");
          //针对每页的HTML,调用Parse类中的方法进行解析
          List<NewsModel> datalist = Parse.getData(html, page);
          //针对已获取的数据,调用MYSQLControl中的方法插入数据
          MYSQLControl.executeInsert(datalist);
      }
   }
 }
```

从主类 CrawlerMain 中可以发现，网络爬虫的整体逻辑，即针对 URL 获取其网页内容，针对网页内容进行解析，最后将解析的数据存放到数据库。运行程序 CrawlerMain，控制台的输出结果如图 7.7 所示。

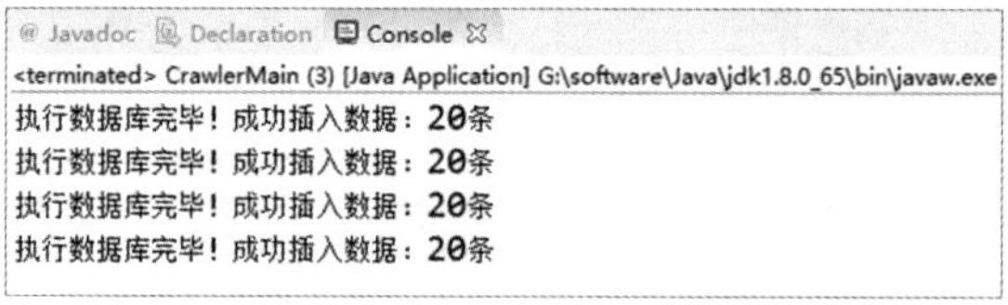

图 7.7 CrawlerMain 对应的控制台的输出结果

在数据库管理工具 Navicat 中，打开 fnlondonnews 表可以发现网络爬虫采集的数据成功插入数据库，如图 7.8 所示。

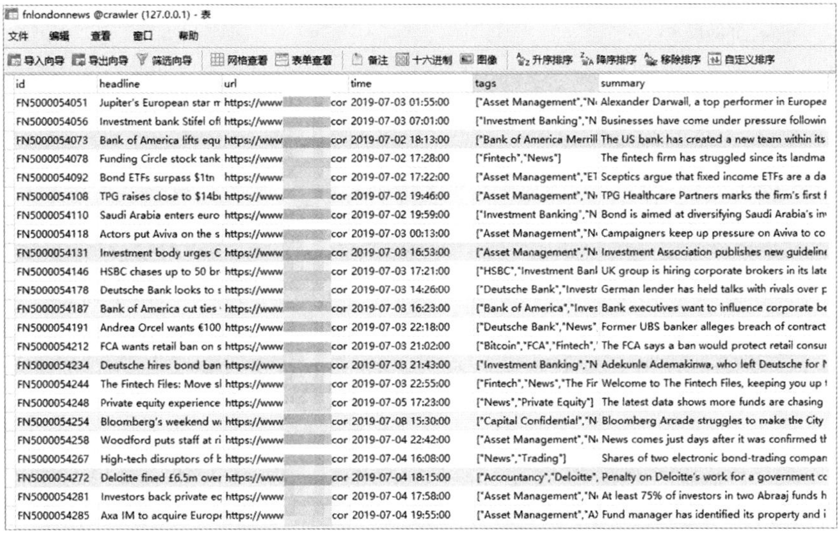

图 7.8　fnlondonnews 表中的数据

7.2　企业信息采集

7.2.1　采集的网页

本项目以采集某集团网站上的企业信息为例。具体包括两层页面，第一层页面采集的是所有企业的基本信息（如企业名称、企业简介、URL 等）；针对第一层页面所采集的数据，获取第二层页面的待采集 URL，进而采集第二层页面的企业详细信息（如企业的详细位置、企业所属国家、企业传真、创建时间等）。图 7.9 所示为第一层页面对应的信息，图 7.10 所示为第二层页面对应的信息。

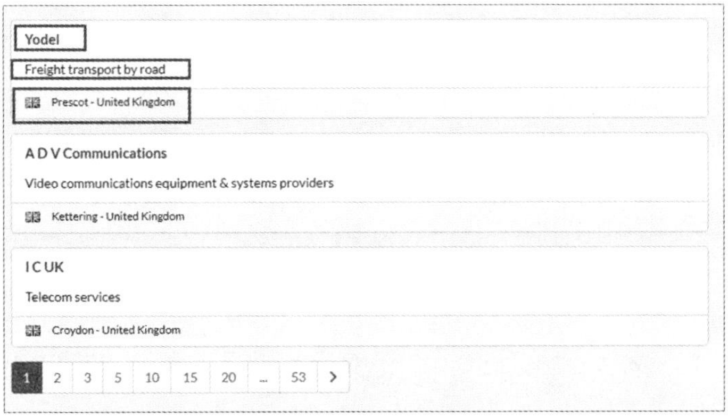

图 7.9　第一层页面对应的信息

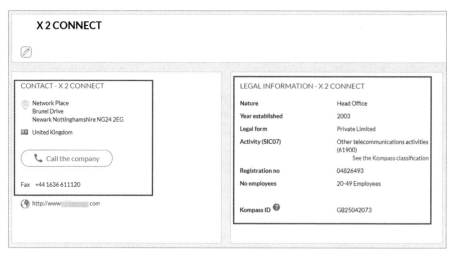

图 7.10　第二层页面对应的信息

第一层页面的信息对应的网页 URL 为 https://gb.*******.com/…。

单击图 7.9 中的翻页按钮，并使用浏览器进行抓包分析，可以得到第二页网页内容对应的真实 URL 为 https://gb.*******.com/…/page-2/。可以发现改变 URL 中的"page"后的数字，可以看到实现翻页操作。

其响应数据为 HTML 格式，为此，需要使用 Jsoup 工具进行解析。其中，获取的第一层页面的信息包括企业 id、企业名称、企业简介、企业所属地域和企业 URL，如图 7.11 所示。

```
<div id="seoProdListGB25080212" class="prod_list ">
    <div class="product-list-data">
        <h2>
            <a id="seoCompanyLinkGB25080212" href="https://gb.      .com/c/yodel/gb25080212/" title="Yodel">
                Yodel</a>
        </h2>
        <div class="row rowTop">
            <div class="col-xs-12 col-sm-12 col-md-12">
                <div class="row rowFlex">
                    <div class="details">
                        <p class="product-summary">
                            <a href="https://gb.      .com/c/yodel/gb25080212/" title="Yodel">Freight transport by road</a>
                        </p>
                    </div>
                </div>
            </div>
        </div>
        <div class="row rowFooter">
            <div class="col-xs-12 col-sm-12 col-md-8 place">
                <a href="https://      .com/c/yodel/gb25080212/" title="Yodel" class="flagWorld">
                    <span class="flag-gb"></span>
                    <span class="placeText">Prescot - United Kingdom</span>
                </a>
            </div>
```

图 7.11　HTML 中的企业基本信息

第二层页面的信息主要有两大类，即企业联系信息和企业相关法律信息。其中，企业联系信息包括企业的详细位置、企业所属国家等；企业法律信息包括企业性质、创建时间等。例如，X2CONNECT 公司的详细信息页面对应的 URL 为 https://gb.*******.com/c/x-2-connect/gb25042073/。

通过网络抓包后发现，后台返回的数据也封装在 HTML 中，图 7.12 和图 7.13 分别展示了 HTML 中包含的企业联系信息和企业相关法律信息。从图 7.13 中可以看到企业性质（Nature）、创建时间（Year established）等字段都存放于<tr>元素中，但这些字段并没有具体属性标记。为此，这里使用判断语句匹配每个字段对应的内容。例如，使用 Jsoup 抽取<tr>元素下的子元素<th>，并判断文本内容是否包含"Nature"，如果包含则抽取该<tr>元素下的子元素<td>，同时获取<td>下的文本内容，作为企业性质（Nature）的字段值。

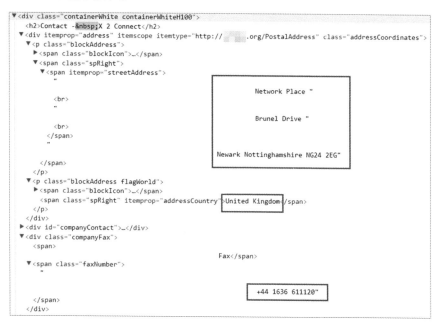

图 7.12　HTML 中的企业联系信息

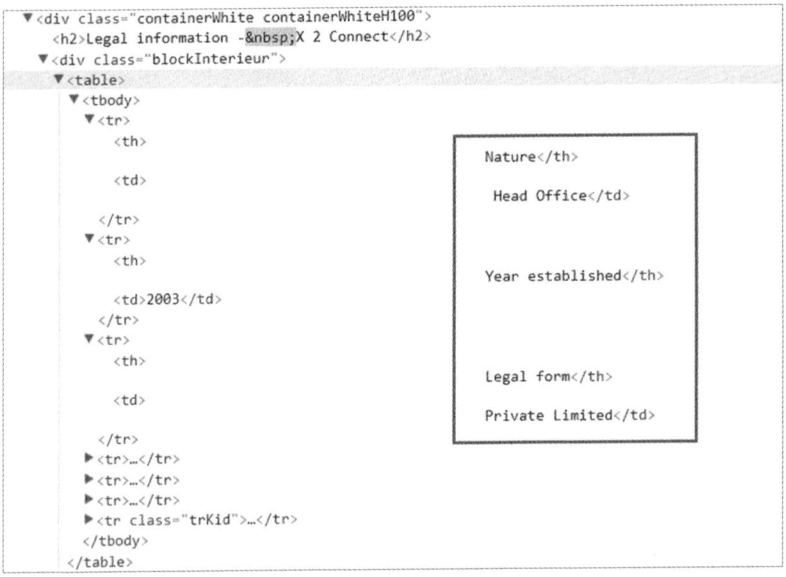

```
▼<div class="containerWhite containerWhiteH100">
    <h2>Legal information - X 2 Connect</h2>
  ▼<div class="blockInterieur">
    ▼<table>
      ▼<tbody>
        ▼<tr>
            <th>                          Nature</th>
            <td>
                                          Head Office</td>
         </tr>
        ▼<tr>
            <th>
                                          Year established</th>
            <td>2003</td>
         </tr>
        ▼<tr>
            <th>
                                          Legal form</th>
            <td>
                                          Private Limited</td>
         </tr>
         ▶<tr>…</tr>
         ▶<tr>…</tr>
         ▶<tr>…</tr>
         ▶<tr class="trKid">…</tr>
         </tbody>
       </table>
```

图 7.13　HTML 中的企业法律信息

7.2.2　框架介绍

本项目的案例对应的程序仍采用 7.1 节的框架结构编写，框架中的 package 有 com.（db、main、model、parse、util）。图 7.14 所示为框架的结构。其中，com.util 中包含了对 HTTPS 处理的文件 SSL509TrustManager.java。

在编写代码之前，需要使用 Maven 工程中的 pom.xml 文件配置整个项目需要的 jar 包。

```xml
<!-- 数据库相关 -->
<dependency>
    <groupId>mysql</groupId>
    <artifactId>mysql-connector-java</artifactId>
    <version>5.1.32</version>
</dependency>
<dependency>
    <groupId>commons-dbutils</groupId>
    <artifactId>commons-dbutils</artifactId>
    <version>1.7</version>
</dependency>
<dependency>
    <groupId>org.apache.commons</groupId>
    <artifactId>commons-dbcp2</artifactId>
```

```xml
        <version>2.5.0</version>
    </dependency>
    <!-- HTML解析 -->
    <dependency>
        <groupId>org.jsoup</groupId>
        <artifactId>jsoup</artifactId>
        <version>1.11.3</version>
    </dependency>
    <!-- 易于处理文件与集合的类库 -->
    <dependency>
        <groupId>com.google.guava</groupId>
        <artifactId>guava</artifactId>
        <version>26.0-jre</version>
    </dependency>
```

图 7.14　框架的结构

7.2.3　第一层信息采集

1．com.model

针对第一层页面，需要采集的数据字段包括企业 id、企业名称、企业简介、所属地域和对应的 URL。为此，创建 BasicInfoModel 类，如程序 7-10 所示。

程序 10

```java
package com.model;
//用于封装企业基本信息,即第一层页面
public class BasicInfoModel {
```

```java
    private String id; //企业id
    private String name;//企业名称
    private String summary; //企业简介
    private String placeText;  //所属地域
    private String url;  //对应的URL
    public String getName() {
        return name;
    }
    public void setName(String name) {
        this.name = name;
    }
    public String getId() {
        return id;
    }
    public void setId(String id) {
        this.id = id;
    }
    public String getSummary() {
        return summary;
    }
    public void setSummary(String summary) {
        this.summary = summary;
    }
    public String getPlaceText() {
        return placeText;
    }
    public void setPlaceText(String placeText) {
        this.placeText = placeText;
    }
    public String getUrl() {
        return url;
    }
    public void setUrl(String url) {
        this.url = url;
    }
}
```

2. com.util

com.util 中存放的是 URL 请求类 HttpRequestUtil 和 SSL509TrustManager。其中，SSL509TrustManager 的功能是认证客户端与服务器，进而确保数据发送到正确的客户端和服务器。在请求一些以 https:// 为前缀的 URL 时，如果没有创建证书信任管理器，则会产生一些错误。SSL509TrustManager 类的内容如程序 7-11 所示。

程序 7-11

```
package com.util;
import javax.net.ssl.X509TrustManager;
import java.security.cert.CertificateException;
import java.security.cert.X509Certificate;
//实现X509TrustManager接口
public class SSL509TrustManager implements X509TrustManager {
  //检查客户端证书
  public void checkClientTrusted(X509Certificate[] x509Certificates,
String s) throws CertificateException {
      //do nothing 接受任意客户端证书
  }
  //检查服务器端证书
  public void checkServerTrusted(X509Certificate[] x509Certificates,
String s) throws CertificateException {
      //do nothing 接受任意服务器端证书
  }
  //返回受信任的X509证书
  public X509Certificate[] getAcceptedIssuers() {
      return new X509Certificate[0];
  }
}
```

HttpRequestUtil 类的内容如程序 7-12 所示。相比于 7.1.3 节中的 HttpRequestUtil，这里多添加了一种 initSSLClient() 方法（已在 4.2.9 节中介绍）。

程序 7-12

```
package com.util;
import org.apache.http.*;
import org.apache.http.client.ClientProtocolException;
import org.apache.http.client.HttpClient;
import org.apache.http.client.config.AuthSchemes;
import org.apache.http.client.config.CookieSpecs;
import org.apache.http.client.config.RequestConfig;
import org.apache.http.client.methods.HttpGet;
import org.apache.http.config.Registry;
import org.apache.http.config.RegistryBuilder;
import org.apache.http.conn.socket.ConnectionSocketFactory;
import org.apache.http.conn.socket.PlainConnectionSocketFactory;
```

```java
import org.apache.http.conn.ssl.NoopHostnameVerifier;
import org.apache.http.conn.ssl.SSLConnectionSocketFactory;
import org.apache.http.impl.client.DefaultHttpRequestRetryHandler;
import org.apache.http.impl.client.HttpClients;
import org.apache.http.impl.conn.PoolingHttpClientConnectionManager;
import org.apache.http.message.BasicHeader;
import org.apache.http.util.EntityUtils;
import javax.net.ssl.*;
import java.io.*;
import java.security.KeyManagementException;
import java.security.NoSuchAlgorithmException;
import java.util.ArrayList;
import java.util.Arrays;
import java.util.List;
public class HttpRequestUtil {
    private HttpClient httpClient;
    private List<Header> headerList = new ArrayList<Header>();
    public HttpEntity getEntityByHttpGetMethod(String url){
        HttpGet httpget = new HttpGet(url); //使用的请求方法
        HttpResponse response = null;
        try {
            response = httpClient.execute(httpget);
        } catch (ClientProtocolException e) {
            e.printStackTrace();
        } catch (IOException e) {
            e.printStackTrace();
        }
        HttpEntity httpEntity = response.getEntity();  //获取网页内容流
        return httpEntity;
    }
    //获取URL对应的网页内容
    public String getHTMLContentByHttpGetMethod(String url,String code){
        try {
            return EntityUtils.toString(getEntityByHttpGetMethod(url),code);
        } catch (IOException e) {
            e.printStackTrace();
            return null;
        }
    }
```

```java
/**
 * 基于SSL配置HttpClient
 * @param SSLProtocolVersion(SSL,SSLv3,TLS,TLSv1,TLSv1.1,TLSv1.2)
 * @return httpClient
 */
public void initSSLClient(){
    RequestConfig defaultConfig = null;
    PoolingHttpClientConnectionManager pcm = null;
    try {
        //创建信任管理
        X509TrustManager xtm = new SSL509TrustManager();
        //创建SSLContext对象，并使用指定的信任管理器初始化
        SSLContext context = SSLContext.getInstance("TLS");
        context.init(null, new X509TrustManager[]{xtm}, null);
        //从SSLContext对象中得到SSLConnectionSocketFactory对象
        SSLConnectionSocketFactory sslConnectionSocketFactory = new
                SSLConnectionSocketFactory(context, NoopHostnameVerifier.INSTANCE);
        //设置全局请求配置
        defaultConfig = RequestConfig.custom().setCookieSpec(CookieSpecs.STANDARD_STRICT)
                .setExpectContinueEnabled(true)
                .setTargetPreferredAuthSchemes(Arrays.asList(AuthSchemes.NTLM, AuthSchemes.DIGEST))
                .setProxyPreferredAuthSchemes(Arrays.asList(AuthSchemes.BASIC)).build();
        //注册Http套接字工厂和Https套接字工厂
        Registry<ConnectionSocketFactory> sfr = RegistryBuilder.<ConnectionSocketFactory>create()
                .register("http", PlainConnectionSocketFactory.INSTANCE)
                .register("https", sslConnectionSocketFactory).build();
        //基于str创建连接管理器
        pcm = new PoolingHttpClientConnectionManager(sfr);
    }catch(NoSuchAlgorithmException | KeyManagementException e){
        e.printStackTrace();
    }
    initDefaultHeaders(); //头信息
    //基于连接管理器和配置，实例化HttpClient
    httpClient = HttpClients.custom().
            setConnectionManager(pcm).
```

```java
            //默认重试次数
            setRetryHandler(new DefaultHttpRequestRetryHandler()).
            setDefaultHeaders(headerList).    //添加头信息
            setDefaultRequestConfig(defaultConfig)
            .build();
}
//头信息设置
private List<Header> initDefaultHeaders(){
    headerList.add(new BasicHeader(HttpHeaders.ACCEPT,
            "text/html,application/xhtml+xml,application/xml;q=0.9," +
                    "image/webp,image/apng,*/*;q=0.8"));
    headerList.add(new BasicHeader(HttpHeaders.USER_AGENT,
            "Mozilla/5.0 (Windows NT 10.0; Win64; x64) " +
                    "AppleWebKit/537.36 (KHTML, like Gecko)"
                    + " Chrome/60.0.3112.113 Safari/537.36"));
    headerList.add(new BasicHeader(HttpHeaders.ACCEPT_ENCODING,
            "gzip, deflate"));
    headerList.add(new BasicHeader(HttpHeaders.CACHE_CONTROL,
            "max-age=0"));
    headerList.add(new BasicHeader(HttpHeaders.CONNECTION,
            "keep-alive"));
    headerList.add(new BasicHeader(HttpHeaders.ACCEPT_LANGUAGE,
            "zh-CN,zh;q=0.8,en;q=0.6,zh-TW;q=0.4,ja;q=0.2," +
                    "de;q=0.2"));
    return headerList;
}
}
```

在请求以 https:// 为前缀的 URL 时,首先需要调用 HttpRequestUtil 类中的 initSSLClient()方法实例化 HttpClient。

3. com.parse

com.parse 中包含一个 Parse 类,负责解析第一层页面及第二层页面的 HTML 数据,这里利用 Jsoup 工具进行解析。针对第一层页面的 HTML 数据,解析代码如程序 7-13 所示。

程序 7-13

```java
//解析第一层页面,获取企业的基本信息
public static List<BasicInfoModel> getBasicData(String html){
    //每页包含多家企业,并封装在集合中
```

```java
        List<BasicInfoModel> list = new ArrayList<BasicInfoModel>();
        //使用Jsoup解析
        Document document = Jsoup.parse(html);
        Elements elements = document.getElementById("resultatDivId")
                .select("div[class~=prod_list?]");
        //循环每家企业
        for (Element ele : elements) {
            String name = ele.select("h2>a").text();
            String id = ele.select("h2>a").attr("id");
            String url = ele.select("h2>a").attr("href");
            //通过class选择
            String summary = ele.select(".details").text();
            String placeText = ele.select(".placeText").text();
            BasicInfoModel model = new BasicInfoModel();
            model.setId(id);
            model.setName(name);
            model.setUrl(url);
            model.setSummary(summary);
            model.setPlaceText(placeText);
            list.add(model);
        }
        return list;
```

4. com.db

com.db 中包含两个 Java 文件，即 MyDataSource 与 MYSQLControl。MyDataSource 类的写法与 7.1.3 节中该类的写法一致。为存储采集的数据，需要在指定数据库(crawler) 中，创建对应的数据表，以下为建表语句：

```sql
CREATE TABLE `kompass_basicinfo` (
  `id` varchar(100) NOT NULL COMMENT '企业id',
  `name` varchar(200) DEFAULT NULL COMMENT '企业名称',
  `summary` varchar(200) DEFAULT NULL COMMENT '企业简介',
  `place` varchar(100) DEFAULT NULL COMMENT '所属地域',
  `url` varchar(100) DEFAULT NULL COMMENT '企业URL',
  PRIMARY KEY (`id`)
) ENGINE=InnoDB DEFAULT CHARSET=utf8;
```

将 BasicInfoModel 实体类对应的数据插入 kompass_basicinfo 表，需要在 MYSQLControl 类中编写程序 7-14 所示的方法。

程序 7-14

```java
    public static void executeInsertBasicInfo(List<BasicInfoModel> list) {
        /** 定义一个Object数组，行列
         * 5表示列数，根据自己的数据定义这里面的数字
         * params[i][0]等是对数组赋值，这里用到集合的GET方法
         **/
        Object[][] params = new Object[list.size()][5];
        for ( int i = 0; i < params.length; i++ ){
          params[i][0] = list.get(i).getId();
          params[i][1] = list.get(i).getName();
          params[i][2] = list.get(i).getSummary();
          params[i][3] = list.get(i).getPlaceText();
          params[i][4] = list.get(i).getUrl();
        }
        try {
          qr.batch("insert into kompass_basicinfo (id,name,"
              + "summary,place,url)"
              + "values (?,?,?,?,?)", params);
        } catch (SQLException e) {
            e.printStackTrace();
        }
        System.out.println("执行数据库完毕！"+"成功插入数据："+ list.size() + "条");
    }
```

5. com.main

在 com.main 中，编写主类 KompassBasicMain。首先，实例化 HttpRequestUtil，并调用 initSSLClient()方法为 HttpClient 配置 SSL，之后请求 URL 获取 HTML 数据，并调用 Parse 类中的 getCodeData()方法解析 HTML 数据，最后调用 MYSQLControl 类中的 executeInsertBasicInfo ()方法将采集的数据插入指定数据表。由于页面存在翻页的情况（见图 7.9），所以这里使用循环的方式拼接每页的 URL。程序 7-15 为主类 KompassBasicMain。

程序 7-15

```java
package com.main;
import java.io.IOException;
import java.util.List;
import com.db.MYSQLControl;
import com.model.BasicInfoModel;
import com.parse.Parse;
import com.util.HttpRequestUtil;
public class KompassBasicMain {
    public static HttpRequestUtil request = new HttpRequestUtil();
    public static void main(String[] args) throws IOException {
        //初始化安全协议，防止不让访问
        request.initSSLClient();
        //请求数据，需要设置页数，这里设置了5页
        for (int i = 1; i < 5; i++) {
            String html = request.getHTMLContentByHttpGetMethod(
                    "https://gb.*******.com/a/postal-services"
                        + "-telecommunications-radio-and"
                        + "-television/79/page-"+ i +"/","UTF-8");
            //解析数据，封装数据
            List<BasicInfoModel> list = Parse.getBasicData(html);
            //存储数据
            MYSQLControl.executeInsertBasicInfo(list);
        }
    }
}
```

执行该程序，控制台的输出结果如图 7.15 所示。

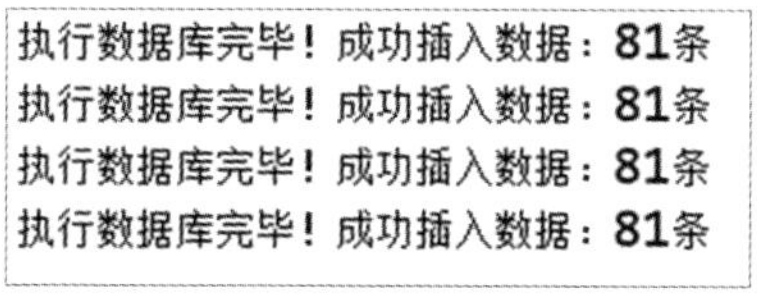

```
执行数据库完毕！成功插入数据：81条
执行数据库完毕！成功插入数据：81条
执行数据库完毕！成功插入数据：81条
执行数据库完毕！成功插入数据：81条
```

图 7.15　KompassBasicMain 对应的控制台的输出结果

在 Navicat 中，打开 kompass_basicinfo 表可以发现网络爬虫采集的数据成功插入了数据库，如图 7.16 所示。

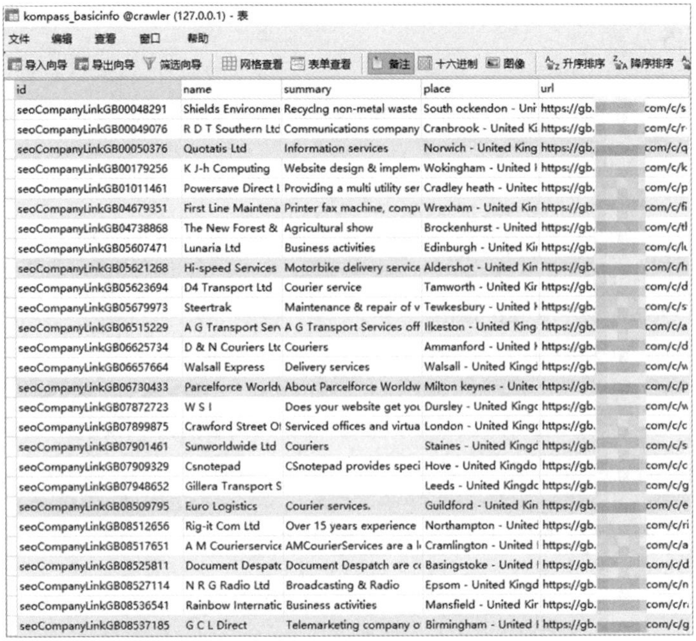

图 7.16　kompass_basicinfo 表中的数据

7.2.4　第二层信息采集

1. com.model

针对第二层页面，需要采集的数据字段包括企业 id、企业的详细位置、企业所属国家、企业传真、企业性质、创建时间、企业类型、经营活动、注册码和员工个数。为此，需要创建 DetailInfoModel 类，内容如程序 7-16 所示。

程序 7-16

```
package com.model;
public class DetailInfoModel {
  private String id;             //企业id
  private String location;       //企业的详细位置
  private String country;        //企业所属国家
  private String fax;            //企业传真
  private String nature;         //企业性质
  private String est_time;       //创建时间
  private String legal_form;     //企业类型
  private String activity;       //经营活动
  private String reg_no;         //注册码
  private String emp_no;         //员工个数
```

```java
    public String getId() {
        return id;
    }
    public void setId(String id) {
        this.id = id;
    }
    public String getLocation() {
        return location;
    }
    public void setLocation(String location) {
        this.location = location;
    }
    public String getCountry() {
        return country;
    }
    public void setCountry(String country) {
        this.country = country;
    }
    public String getFax() {
        return fax;
    }
    public void setFax(String fax) {
        this.fax = fax;
    }
    public String getNature() {
        return nature;
    }
    public void setNature(String nature) {
        this.nature = nature;
    }
    public String getEst_time() {
        return est_time;
    }
    public void setEst_time(String est_time) {
        this.est_time = est_time;
    }
    public String getLegal_form() {
        return legal_form;
    }
    public void setLegal_form(String legal_form) {
        this.legal_form = legal_form;
    }
```

```java
    public String getActivity() {
        return activity;
    }
    public void setActivity(String activity) {
        this.activity = activity;
    }
    public String getReg_no() {
        return reg_no;
    }
    public void setReg_no(String reg_no) {
        this.reg_no = reg_no;
    }
    public String getEmp_no() {
        return emp_no;
    }
    public void setEmp_no(String emp_no) {
        this.emp_no = emp_no;
    }
}
```

2. com.parse

为解析第二层页面的数据，需要在 Parse 类中，再添加一种 getBasicInfoData()方法，如程序 7-17 所示。该方法的输入参数为企业 id 以及该企业页面的 HTML。另外，在该方法中使用了判断语句匹配企业性质、创建时间等字段。

程序 7-17

```java
    //解析每个企业的详细信息
    public static DetailInfoModel getDetailInfo(String id, String html){
        DetailInfoModel model = new DetailInfoModel();
        Document doc = Jsoup.parse(html);
        //解析联系信息
        String location = doc.select("p[class=blockAddress]").text();
        String country = doc.select("p[class=blockAddress flagWorld]").text();
        String fax = doc.select("span[class=faxNumber]").text();
        //解析公司的法律信息
        String nature = "";
        String est_time = "";
        String legal_form = "";
```

```java
        String activity = "";
        String reg_no = "";
        String emp_no = "";
        Elements elements2 = doc.select("div[class=blockInterieur]")
                .select("tr");
//使用判断语句进行匹配
        for (Element ele : elements2) {
            String text = ele.select("th").text();
            if (text.contains("Nature")) {
                nature = ele.select("td").text();
            }else if (text.contains("established")) {
                est_time = ele.select("td").text();
            }else if (text.contains("form")) {
                legal_form = ele.select("td").text();
            }else if (text.contains("Activity")) {
                activity = ele.select("td").text();
            }else if (text.contains("Registration")) {
                reg_no = ele.select("td").text();
            }else if (text.contains("employees")) {
                emp_no = ele.select("td").text();
            }
        }
//封装到对象中
        model.setId(id);
        model.setLocation(location);
        model.setCountry(country);
        model.setFax(fax);
        model.setNature(nature);
        model.setEst_time(est_time);
        model.setLegal_form(legal_form);
        model.setActivity(activity);
        model.setReg_no(reg_no);
        model.setEmp_no(emp_no);
        return model;
    }
```

3. com.db

针对 com.model 中的 DetailInfoModel，需要在数据库中构建对应的数据表，建表语句如下所示。

```sql
CREATE TABLE `kompass_detailinfo` (
  `id` varchar(100) NOT NULL COMMENT '企业id',
  `location` varchar(200) DEFAULT NULL COMMENT '企业的详细位置',
  `country` varchar(100) DEFAULT NULL COMMENT '企业所属国家',
  `fax` varchar(100) DEFAULT NULL COMMENT '企业传真',
  `nature` varchar(50) DEFAULT NULL COMMENT '企业性质',
  `est_time` varchar(50) DEFAULT NULL COMMENT '创建时间',
  `legal_form` varchar(50) DEFAULT NULL COMMENT '企业类型',
  `activity` varchar(200) DEFAULT NULL COMMENT '经营活动',
  `reg_no` varchar(100) DEFAULT NULL COMMENT '注册码',
  `emp_no` varchar(100) DEFAULT NULL COMMENT '员工个数',
  PRIMARY KEY (`id`)
) ENGINE=InnoDB DEFAULT CHARSET=utf8;
```

在 MYSQLControl 类中添加将数据插入数据表 kompass_detailinfo 的方法，如程序 7-18 所示。

程序 7-18

```java
public static void executeInsertDetailInfo(List<DetailInfoModel> list) {

    /** 定义一个Object数组，行列
     * 10表示列数，根据自己的数据定义这里面的数字
     * params[i][0]等是对数组赋值，这里用到集合的GET方法
     **/

    Object[][] params = new Object[list.size()][10];
    for ( int i=0; i<params.length; i++ ){
      params[i][0] = list.get(i).getId();
      params[i][1] = list.get(i).getLocation();
      params[i][2] = list.get(i).getCountry();
      params[i][3] = list.get(i).getFax();
      params[i][4] = list.get(i).getNature();
      params[i][5] = list.get(i).getEst_time();
      params[i][6] = list.get(i).getLegal_form();
      params[i][7] = list.get(i).getActivity();
      params[i][8] = list.get(i).getReg_no();
      params[i][9] = list.get(i).getEmp_no();
    }
    try {
        qr.batch("insert into kompass_detailinfo "
            + "values (?,?,?,?,?,?,?,?,?,?)", params);
```

```
        } catch (SQLException e) {
            e.printStackTrace();
        }
        System.out.println("执行数据库完毕！" + "成功插入数据: " + list.size() + "条");    }
```

由于需要将第一层获取的所有企业 id 和企业 URL 数据从 kompass_basicinfo 表中抽取出来，所以在 MYSQLControl 类中还要添加一种抽取数据的方法，如程序 7-19 所示。

程序 7-19

```
        public static <T> List<T> getListInfoBySQL (String sql, Class<T> type ){
            List<T> list = null;
            try {
                list = qr.query(sql,new BeanListHandler<T>(type));
            } catch (SQLException e) {
                e.printStackTrace();
            }
            return list;
        }
```

4．com.main

在 com.main 中，编写主类 KompassDetailInfoMain。首先，实例化 HttpRequestUtil，并调用 initSSLClient()方法为 HttpClient 配置 SSL，之后从数据库中抽取 id 和 URL 字段（调用 MYSQLControl 类中的 getListInfoBySQL 方法），并封装到集合中。

这里使用队列存储 BasicInfoModel 对象（包含 id 和 URL 字段），循环队列中的每个 BasicInfoModel 对象，获取 URL，进而执行数据采集操作。如果频繁请求此网站，可能面临 ip 被封的危险，因此要在主方法中添加随机休息时间，即每采集一定数量的 URL，休息一段时间。另外，如果出现断网或者请求的数据为 null 的情况，表示某 BasicInfoModel 对象对应的 URL 采集失败，为此需要将此次的 BasicInfoModel 对象重新添加到队列中。再者，Parse 类中的 getDetailInfo()方法解析的数据封装在 DetailInfoModel 对象中，为实现批量存储的目的，这里又将每个对象的数据封装于集合中（listData）。

程序 7-20 所示为主类 KompassDetailInfoMain。执行该程序，对应控制台的输出结果如图 7.17 所示。

程序 7-20

```java
package com.main;
import java.util.ArrayList;
import java.util.LinkedList;
import java.util.List;
import java.util.Queue;
import com.db.MYSQLControl;
import com.model.DetailInfoModel;
import com.model.BasicInfoModel;
import com.parse.Parse;
import com.util.HttpRequestUtil;
public class KompassDetailInfoMain {
    public static HttpRequestUtil request = new HttpRequestUtil();
    public static void main(String[] args) throws InterruptedException {
        //初始化安全协议,防止不让访问
        request.initSSLClient();
        //从数据库取数据
        List<BasicInfoModel> basicList = MYSQLControl
    .getListInfoBySQL("select id,url from "
                + "kompass_basicinfo where id "
                + "NOT IN (select id from kompass_detailinfo)",
                BasicInfoModel.class);
        //待爬的队列
        Queue<BasicInfoModel> queue =
    new LinkedList<BasicInfoModel>();
        for( int i = 0; i < basicList.size(); i++ ){
            queue.offer(basicList.get(i));
        }
        boolean t = true;
        int count = 1;
        List<DetailInfoModel> listData = new ArrayList
<DetailInfoModel>();
        while (t) {
            //如果队列为空,循环结束
            if( queue.isEmpty() ){
                t = false;
            }else {
                //待爬的某个企业
                BasicInfoModel oneCompany = queue.poll();
                //获取HTML
                String html = request.getHTMLContentByHttpGetMethod(
```

```
        oneCompany.getUrl(),"utf-8");
                //为防止断网或其他因素
                if (!html.contains("Nature")||(html == null)) {
                    queue.add(oneCompany);
                }else {
                    //解析数据
                    DetailInfoModel model = Parse
                        .getDetailInfo(oneCompany.getId(), html);
                    //封装到集合
                    listData.add(model);
                }
                //批量存储数据
                if (listData.size() == 3) {
    MYSQLControl.executeInsertDetailInfo(listData);
                    listData.clear();
                }
                if (count%3 == 0) {
                    //产生随机数
                    int m = (int)(Math.random()*2) + 5;
                    Thread.sleep(m*1000);
                }
                count++;
            }
        }
        //剩余数据统一插入数据库
        MYSQLControl.executeInsertDetailInfo(listData);
    }
}
```

图 7.17　KompassDetailInfoMain 对应控制台的输出结果

在 Navicat 中，打开 kompass_detailinfo 表可以发现网络爬虫采集的数据成功插入了数据库，如图 7.18 所示。

kompass_detailinfo @crawler (127.0.0.1) - 表

文件 编辑 查看 窗口 帮助

导入向导 导出向导 筛选向导 网格查看 表单查看 备注 十六进制 图像 升序排序 降序排序 移除排序 自定义排序

id	location	country	fax	nature	est_time	legal_form	activity	reg_no	emp_no
seoCompanyLinkGB000482	Kerry Avenue Aveley South Ockendon Ess	United Kingdom	+44 1708 684020	Head Office	1979	P.L.C.	Recovery of sorted mat	01429915	50-99 Employees
seoCompanyLinkGB000490	Crown Yard, Bedgebury Road Goudhurst	United Kingdom	+44 871 989 8158	Head Office	1999	Private Limitec	Other telecommunicatio	03839170	10-19 Employees
seoCompanyLinkGB000503	Suite 1 Joseph King House, Abbey Farm I	United Kingdom	+44 1603 899919	Head Office	2005	Private Limitec	Other business support	05643725	20-49 Employees
seoCompanyLinkGB001792	19 Wildcroft Drive Wokingham Berkshire	United Kingdom	+44 118 977 5957	Head Office			Repair of computers an		0-9 Employees
seoCompanyLinkGB010114	Cabrhilla House 87-89 Reddal Hill Road C	United Kingdom	+44 870 787 4928	Head Office	2000	Private Limitec	Construction of other ci	03934964	10-19 Employees
seoCompanyLinkGB046793	Bersham Road Bersham Wrexham Clwyd	United Kingdom	+44 1244 289713	Head Office	2002	Private Limitec	Specialised construction	04519502	0-9 Employees
seoCompanyLinkGB047388	The Showground New Park Brockenhurst	United Kingdom	+44 1590 622637	Head Office	1991	Private Limitec	Other business support	02646090	0-9 Employees
seoCompanyLinkGB056074	1333 Shandon Place Edinburgh Midlothiai	United Kingdom	+44 131 443 3712	Head Office		Private Limitec	Other business support	02488424	0-9 Employees
seoCompanyLinkGB056212	Lysons Avenue Ash Vale Aldershot Hamps	United Kingdom	+44 1252 324166	Head Office	1999	Private Limitec	Freight transport by roa	03695819	20-49 Employees
seoCompanyLinkGB056236	55 Ivy Croft Road Warton Tamworth Staff	United Kingdom	+44 1827 897722	Head Office	2004	Private Limitec	Freight transport by roa	05286805	0-9 Employees
seoCompanyLinkGB056799	Unit 1a Commercial House Station Road 1	United Kingdom	+44 1684 276500	Head Office	2005		Other business support	05445776	0-9 Employees
seoCompanyLinkGB065152	26 Mill Acre Close Ilkeston Derbyshire DE	United Kingdom		Head Office		Private Limitec	(null) See the Kompass (	06661758	0-9 Employees
seoCompanyLinkGB066257	Clawdd Du, Pontardulais Road Tycroes Ar	United Kingdom	+44 870 4865401	Head Office	2009	Private Limitec	Freight transport by roa	06907661	0-9 Employees
seoCompanyLinkGB066576	Unit 6 Rollingmill Business Park Rollingmil	United Kingdom		Head Office	2016		Freight transport by roa	10321504	0-9 Employees
seoCompanyLinkGB067304	25 Caldecotte Lake Drive Caldecotte Miltc	United Kingdom	+44 1322 273676	Head Office			Other business support		140 Employees
seoCompanyLinkGB078727	First Floor 76-78 Parsonage Street Dursle	United Kingdom	+44 1453 543859	Head Office			Management consultanc		0-9 Employees
seoCompanyLinkGB078998	83-87 Crawford Street London Greater Lo	United Kingdom	+44 207 616 1701	Head Office	2009		Other business support	05133340	20-49 Employees
seoCompanyLinkGB079014	The Mill Horton Road Staines Middlesex	United Kingdom	+44 1753 680225	Head Office	2003	Private Limitec	Freight transport by roa	04794977	0-9 Employees
seoCompanyLinkGB079093	Gemini Business Centre 136-140 Old Shor	United Kingdom	+44 845 199 9940	Head Office	2013		Other business support	06784130	10-19 Employees
seoCompanyLinkGB079486	4 Armley Grange Drive Leeds West Yorksl	United Kingdom		Head Office	2008	Private Limitec	(null) See the Kompass (	06517127	0-9 Employees
seoCompanyLinkGB085097	2 Emmanuel Close Guildford Surrey GU2	United Kingdom	+44 1483 859027	Head Office	2006		Freight transport by roa	05756083	0-9 Employees
seoCompanyLinkGB085126	Truffell House, High Street Scaldwell Nort	United Kingdom	+44 1604 883570	Head Office	1999	Private Limitec	Other telecommunicatio	03851351	10-19 Employees
seoCompanyLinkGB085176	26 Easington Avenue Hartford Green Crar	United Kingdom	+44 1670 735667	Head Office	2005		Freight transport by roa	06564271	0-9 Employees

图 7.18 kompass_detailinfo 表中的数据

7.3 股票信息采集

7.3.1 采集的网页

本项目以采集某网站汽车板块相关股票的数据为例。

所涉及网址对应的内容如图 7.19 所示。

序号	代码	名称	相关链接	最新价	涨跌幅	涨跌额	成交量(手)	成交额	振幅	最高	最低	今开	昨收	量比	换手率	市盈率(动态)	加自选
1	603023		股吧 资金流 数据	6.13	10.05%	0.56	7.49万	4526万	5.92%	6.13	5.80	5.94	5.57	21.47	2.08%	45.72	+
2	002488		股吧 资金流 数据	9.47	9.99%	0.86	2.89万	2735万	0.00%	9.47	9.47	9.47	8.61	1.14	0.38%	42.26	+
3	600469		股吧 资金流 数据	4.96	9.98%	0.45	5.61万	2714万	11.31%	4.96	4.45	4.46	4.51	12.86	1.00%	1118.83	+
4	603239		股吧 资金流 数据	14.00	9.12%	1.17	4.93万	6719万	8.57%	14.11	13.01	13.01	12.83	10.54	5.44%	24.97	+
5	300176		股吧 资金流 数据	47.08	6.42%	2.84	2.46万	1.15亿	7.41%	48.28	45.00	45.00	44.24	10.30	0.66%	37.41	+
6	603197		股吧 资金流 数据	25.78	5.31%	1.30	9591	2448万	5.68%	25.97	24.58	24.99	24.48	3.37	0.97%	26.29	+
7	601689		股吧 资金流 数据	20.83	4.62%	0.92	4.82万	9897万	6.98%	21.20	19.81	19.91	19.91	3.73	1.97%	18.42	+
8	002239		股吧 资金流 数据	2.64	3.94%	0.10	35.24万	9230万	4.72%	2.67	2.55	2.58	2.54	3.86	1.16%	16.15	+
9	002708		股吧 资金流 数据	9.54	3.14%	0.29	7.04万	6680万	2.38%	9.61	9.39	9.40	9.25	2.18	1.89%	-	+
10	603006		股吧 资金流 数据	12.52	2.88%	0.35	1.34万	1659万	3.12%	12.60	12.22	12.22	12.17	2.62	1.34%	27.30	+
11	603305		股吧 资金流 数据	35.34	2.29%	0.79	8576	3022万	4.86%	35.98	34.30	34.49	34.55	1.54	1.36%	46.18	+
12	603788		股吧 资金流 数据	16.98	2.29%	0.38	1.1万	1866万	2.59%	17.12	16.69	16.79	16.60	3.63	0.48%	14.21	+
13	002031		股吧 资金流 数据	2.51	2.03%	0.05	14.89万	3709万	5.28%	2.55	2.42	2.46	2.46	2.93	0.79%	73.65	+
14	002920		股吧 资金流 数据	26.66	1.91%	0.50	5815	1529万	3.52%	26.66	25.74	26.15	26.16	1.53	0.48%	26.49	+
15	603089		股吧 资金流 数据	19.89	1.84%	0.36	5394	1072万	2.87%	20.11	19.55	19.71	19.53	2.69	1.71%	25.03	+
16	300201		股吧 资金流 数据	4.97	1.84%	0.09	3.86万	1914万	2.87%	5.00	4.86	4.91	4.88	2.58	0.44%	38.59	+
17	600148		股吧 资金流 数据	16.85	1.75%	0.29	8883	1482万	3.02%	16.94	16.44	16.56	16.56	2.03	0.63%	74.04	+
18	000030		股吧 资金流 数据	4.78	1.70%	0.08	2.75万	1307万	2.77%	4.79	4.66	4.70	4.70	2.29	0.16%	9.82	+
19	002592		股吧 资金流 数据	20.12	1.62%	0.32	7710	1546万	2.07%	20.22	19.81	20.00	19.80	1.49	0.31%	239.73	+
20	603926		股吧 资金流 数据	16.06	1.58%	0.25	2857	456.6万	2.15%	16.16	15.82	15.96	15.81	2.63	0.44%	23.25	+

上一页 1 2 3 4 5 … 8 下一页 转到 1 GO

图 7.19 该网站汽车板块股票的数据

首先，使用谷歌浏览器对第一页面的内容进行抓包，抓包结果如图 7.20 所示。从图 7.20 中可以发现该网页对应的信息存放在 JSON 字符串中，其 URL 的形式为 http://nufm.*****.com/…&p=1…。

图 7.20　网络抓包结果

通过对其他页面的内容抓包会发现，调整上面 URL 中的 "p" 参数的值，可以实现翻页的目的。

7.3.2　框架介绍

图 7.21 所示为本项目的案例使用的框架结构，相比于前两节，这里添加了一个新的 package，即 com.job，其功能是负责任务调度。

图 7.21　框架的结构

在编写代码之前，需要使用 Maven 工程中的 pom.xml 文件配置整个项目需要的 jar 包。

```
<!--- httpclient 相关依赖 jar 包 -->
```

```xml
<dependency>
    <groupId>org.apache.httpcomponents</groupId>
    <artifactId>httpclient</artifactId>
    <version>4.5.5</version>
</dependency>
<!-- 数据库相关 -->
<dependency>
    <groupId>mysql</groupId>
    <artifactId>mysql-connector-java</artifactId>
    <version>5.1.32</version>
</dependency>
<dependency>
    <groupId>commons-dbutils</groupId>
    <artifactId>commons-dbutils</artifactId>
    <version>1.7</version>
</dependency>
<dependency>
    <groupId>org.apache.commons</groupId>
    <artifactId>commons-dbcp2</artifactId>
    <version>2.5.0</version>
</dependency>
<!--任务调度-->
<dependency>
    <groupId>org.quartz-scheduler</groupId>
    <artifactId>quartz</artifactId>
    <version>2.3.0</version>
</dependency>
```

7.3.3 程序设计

1. com.model

由图 7.19 可知，本案例需要采集数据的字段包括日期、股票代码、股票名称、最新价、涨跌额、涨跌幅、振幅、成交量（单位：手）、成交额、昨收、今开、最低、最高、股票数据更新时间和数据采集时间。基于这些字段，创建 CarStockModel 类，内容如程序 7-21 所示。

程序 7-21

```java
package com.model;
public class CarStockModel {
```

```java
    private String date;                                  //日期
    private String stock_id;                              //股票代码
    private String stock_name;                            //股票名称
    private float stock_price;                            //最新价
    private float stock_change;                           //涨跌额
    private float stock_range;                            //涨跌幅
    private float stock_amplitude;                        //振幅
    private int stock_trading_number;                     //成交量
    private int stock_trading_value;                      //成交额
    private float stock_yesterdayfinish_price;            //昨收
    private float stock_todaystart_price;                 //今开
    private float stock_max_price;                        //最低
    private float stock_min_price;                        //最高
    private String stock_time;                            //股票数据更新时间
    private String craw_time;                             //数据采集时间
    public String getDate() {
        return date;
    }
    public void setDate(String date) {
        this.date = date;
    }
    public String getStock_id() {
        return stock_id;
    }
    public void setStock_id(String stock_id) {
        this.stock_id = stock_id;
    }
    public String getStock_name() {
        return stock_name;
    }
    public void setStock_name(String stock_name) {
        this.stock_name = stock_name;
    }
    public float getStock_price() {
        return stock_price;
    }
    public void setStock_price(float stock_price) {
        this.stock_price = stock_price;
    }
    public float getStock_change() {
        return stock_change;
    }
```

```java
    public void setStock_change(float stock_change) {
        this.stock_change = stock_change;
    }
    public float getStock_range() {
        return stock_range;
    }
    public void setStock_range(float stock_range) {
        this.stock_range = stock_range;
    }
    public float getStock_amplitude() {
        return stock_amplitude;
    }
    public void setStock_amplitude(float stock_amplitude) {
        this.stock_amplitude = stock_amplitude;
    }
    public int getStock_trading_number() {
        return stock_trading_number;
    }
    public void setStock_trading_number(int stock_trading_number) {
        this.stock_trading_number = stock_trading_number;
    }
    public int getStock_trading_value() {
        return stock_trading_value;
    }
    public void setStock_trading_value(int stock_trading_value) {
        this.stock_trading_value = stock_trading_value;
    }
    public float getStock_yesterdayfinish_price() {
        return stock_yesterdayfinish_price;
    }
    public void setStock_yesterdayfinish_price(float stock_yesterdayfinish_price) {
        this.stock_yesterdayfinish_price = stock_yesterdayfinish_price;
    }
    public float getStock_todaystart_price() {
        return stock_todaystart_price;
    }
    public void setStock_todaystart_price(float stock_todaystart_price) {
        this.stock_todaystart_price = stock_todaystart_price;
    }
    public float getStock_max_price() {
        return stock_max_price;
```

```java
    }
    public void setStock_max_price(float stock_max_price) {
        this.stock_max_price = stock_max_price;
    }
    public float getStock_min_price() {
        return stock_min_price;
    }
    public void setStock_min_price(float stock_min_price) {
        this.stock_min_price = stock_min_price;
    }
    public String getStock_time() {
        return stock_time;
    }
    public void setStock_time(String stock_time) {
        this.stock_time = stock_time;
    }
    public String getCraw_time() {
        return craw_time;
    }
    public void setCraw_time(String craw_time) {
        this.craw_time = craw_time;
    }
}
```

2. com.util

HttpRequestUtil 类的编写与 7.1 节的案例完全相同，这里不再赘述。由于需要存储当天日期 date（格式为 yyyy-MM-dd）和股票数据采集时间 craw_time（格式为 yyyy-MM-dd HH:mm:ss），因此本项目在 TimeUtil 编写了一种获取当前时间的方法，如程序 7-22 所示。

程序 7-22

```java
//获取当前时间
public static String GetNowDate(String formate){
    String temp_str = "";
    Date dt = new Date();
    SimpleDateFormat sdf = new SimpleDateFormat(formate);
    temp_str = sdf.format(dt);
    return temp_str;
}
```

3. com.parse

图 7.22 展示了 Parse 类需要解析的部分 JSON 数据。由图 7.22 可知，不同股票之间的数据以引号和逗号分隔，同一只股票不同字段之间以逗号分隔。

```
["2,002031,▒▒▒▒,
2.71,0.25,10.16,1058868,273167408,11.79,2.71,2.42,2.46,2.46,0.00,1.91,5.60,79.52,1.95,596
0362350,5123568955,22.6%,35.64%,-,2004-08-16,2019-03-05 15:00:00,1058868",
"1,603023,▒▒▒▒,
6.13,0.56,10.05,98839,59920725,5.92,6.13,5.80,5.94,5.57,0.00,2.83,2.75,45.72,3.92,2206800
041,2206800041,14.58%,30.7%,-,2015-05-27,2019-03-05 15:00:00,98839"
]
```

图 7.22　部分 JSON 数据

针对以上格式的 JSON 数据，可使用 split(regex)方法对原字符串进行切分，获取所要采集的字段，详细的解析程序如 7-23 所示。

程序 7-23

```java
package com.parse;
import java.util.ArrayList;
import java.util.List;
import com.model.CarStockModel;
import com.util.TimeUtil;
public class Parse {
    public static List<CarStockModel> getData(String content) throws Exception {
        List<CarStockModel> list = new ArrayList<CarStockModel>();
        //数据预处理
        content = content.split("data:")[1].split (",recordsFiltered")[0];
        String stocks[] = content.split("\",");
        List<String> stocklist = new ArrayList<String>();
        for (int i = 0; i < stocks.length; i++) {
            stocklist.add(stocks[i].replace("[\"", "").replace("\"", "").replace("]", ""));
        }
        //获取数据
        for (int i = 0; i < stocklist.size(); i++) {
            String date = TimeUtil.GetNowDate("yyyy-MM-dd");
            String stock_id = stocklist.get(i).split(",")[1];
            String stock_name = stocklist.get(i).split(",")[2];
```

```java
            //股票价格
            float stock_price=Float.parseFloat
(stocklist.get(i).split(",")[3]);
            //涨跌额
            float stock_change = Float.parseFloat(stocklist.get(i).
split(",")[4]);
            //涨跌幅
            float stock_range = Float.parseFloat(stocklist.get(i).
split(",")[5]);
            //成交量
            int stock_trading_number =
Integer.parseInt(stocklist.get(i).split(",")[6]);
            //成交额
            int stock_trading_value = Integer.parseInt(stocklist.get
(i).split(",")[7]);
            //振幅
            float stock_amplitude = Float.parseFloat(stocklist.get
(i).split(",")[8]);
            //最高
            float stock_max_price = Float.parseFloat(stocklist.get
(i).split(",")[9]);
            //最低
            float stock_min_price = Float.parseFloat(stocklist.get
(i).split(",")[10]);
            //今开
            float stock_todaystart_price =
Float.parseFloat(stocklist.get(i).split(",")[11]);
            //昨收
            float stock_yesterdayfinish_price = Float.parseFloat
(stocklist.get(i).split(",")[12]);
            //股票数据更新时间
            String stock_time = stocklist.get(i).split(",")[24];
            String craw_time = TimeUtil.GetNowDate("yyyy-MM-dd
HH:mm:ss");
            //封装数据
            CarStockModel model = new CarStockModel();
            model.setDate(date);
            model.setStock_id(stock_id);
            model.setStock_name(stock_name);
            model.setStock_price(stock_price);
            model.setStock_change(stock_change);
```

```
                model.setStock_range(stock_range);
                model.setStock_amplitude(stock_amplitude);
                model.setStock_trading_number(stock_trading_number);
                model.setStock_trading_value(stock_trading_value);
                model.setStock_yesterdayfinish_price(stock_
yesterdayfinish_price);
                model.setStock_todaystart_price
(stock_todaystart_price);
                model.setStock_max_price(stock_max_price);
                model.setStock_min_price(stock_min_price);
                model.setStock_time(stock_time);
                model.setCraw_time(craw_time);
                list.add(model);
            }
            return list;
        }
    }
```

4. com.db

基于 CarStockModel 类中的字段，在数据库中构建对应的数据表，建表语句如下所示。

```
CREATE TABLE `dongfang_car_stock` (
  `date` date NOT NULL COMMENT '当天日期',
  `stock_id` char(20) NOT NULL COMMENT '股票代码',
  `stock_name` char(50) DEFAULT NULL COMMENT '股票名称',
  `stock_price` float(10,2) DEFAULT NULL COMMENT '股票最新价格',
  `stock_change` float(10,2) DEFAULT NULL COMMENT '涨跌额',
  `stock_range` float(10,4) DEFAULT NULL COMMENT '涨跌幅度',
  `stock_amplitude` float(10,4) DEFAULT NULL COMMENT '振幅',
  `stock_trading_number` int(20) DEFAULT NULL COMMENT '成交量',
  `stock_trading_value` int(20) DEFAULT NULL COMMENT '成交额',
  `stock_yesterdayfinish_price` float(10,3) DEFAULT NULL COMMENT '昨天收益价格',
  `stock_todaystart_price` float(10,3) DEFAULT NULL COMMENT '今日开盘价格',
  `stock_max_price` float(10,3) DEFAULT NULL COMMENT '股票最高价格',
  `stock_min_price` float(10,3) DEFAULT NULL COMMENT '股票最低价格',
  `stock_time` varchar(40) NOT NULL COMMENT '股票数据更新时间',
  `craw_time` datetime DEFAULT NULL COMMENT '爬取时间',
  PRIMARY KEY (`stock_id`,`stock_time`)
) ENGINE=InnoDB DEFAULT CHARSET=utf8;
```

在 MYSQLControl 类中添加将数据插入数据表 dongfang_car_stock 的方法，如程序 7-24 所示。

程序 7-24

```java
//将采集的数据插入数据表
    public static void insertCarStocks ( List<CarStockModel> carstocks ) {
        Object[][] params = new Object[carstocks.size()][15];
        for ( int i = 0; i < carstocks.size(); i++ ){
            params[i][0] = carstocks.get(i).getDate();
            params[i][1] = carstocks.get(i).getStock_id();
            params[i][2] = carstocks.get(i).getStock_name();
            params[i][3] = carstocks.get(i).getStock_price();
            params[i][4] = carstocks.get(i).getStock_change();
            params[i][5] = carstocks.get(i).getStock_range();
            params[i][6] = carstocks.get(i).getStock_amplitude();
            params[i][7] = carstocks.get(i).getStock_trading_number();
            params[i][8] = carstocks.get(i).getStock_trading_value();
            params[i][9] = carstocks.get(i).getStock_yesterdayfinish_price();
            params[i][10] = carstocks.get(i).getStock_todaystart_price();
            params[i][11] = carstocks.get(i).getStock_max_price();
            params[i][12] = carstocks.get(i).getStock_min_price();
            params[i][13] = carstocks.get(i).getStock_time();
            params[i][14] = carstocks.get(i).getCraw_time();
        }
        QueryRunner qr = new QueryRunner(ds);
        try {
            qr.batch("INSERT INTO `dongfang_car_stock` VALUES (?,?,?,"
                    + "?,?,?,?,?,?,?,?,?,?,?,?)", params);
            System.out.println("执行数据库完毕！" + "成功插入数据：" + carstocks.size() + "条");
        } catch (SQLException e) {
            System.out.println(e);
        }
    }
```

5. com.main

在 com.main 中，编写主类 CrawlerMain，如程序 7-25 所示。由于汽车相关的股票共有 8 页，所以程序 7-25 中的 i 的值需要从 1 循环到 8。

程序 7-25

```java
package com.main;
import java.util.List;
import com.db.MYSQLControl;
import com.model.CarStockModel;
import com.parse.Parse;
import com.util.HttpRequestUtil;
public class CrawlerMain {
    static HttpRequestUtil httpRequest = new HttpRequestUtil();
    public static void main(String[] args) throws Exception {
        for (int i = 1; i < 9; i++) {
            //拼接URL
            String url = "http://nufm.*****.com/EM_Finance2014"
                + "NumericApplication/JS.aspx?type=CT&token="
                + "4f1862fc3b5e77c150a2b985b12db0fd&sty="
                + "FCOIATC&js=(%7Bdata%3A%5B(x)%5D%2Crecords"
                + "Filtered%3A(tot)%7D)&cmd="
                + "C.BK04811&st=(ChangePercent)&sr=-1&p="
                + i + "&ps=20&_=1551750725008";
            //请求拼接后的URL
            String content = httpRequest.getContentByHttpGetMethod(url,"utf-8");
            //解析URL
            List<CarStockModel> carstocks = Parse.getData(content);
            //存储数据
            MYSQLControl.insertCarStocks(carstocks);
        }
    }
}
```

执行该程序，对应控制台的输出结果如图 7.23 所示。

图 7.23 对应控制台的输出结果

在 Navicat 中，打开 dongfang_car_stock 表可以发现网络爬虫采集的数据成功插入了数据库，如图 7.24 所示。

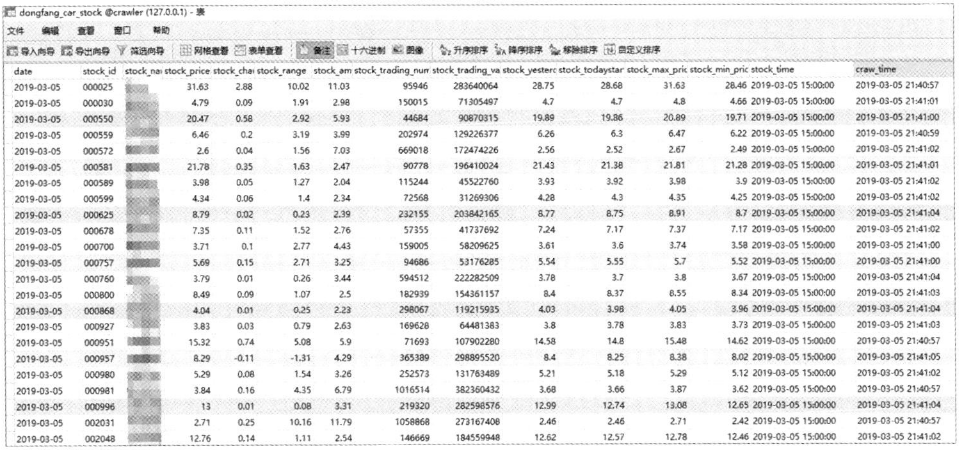

图 7.24　dongfang_car_stock 表中的数据

7.3.4　Quartz 实现定时调度任务

正常情况下，沪深市场股票的交易时间是周一到周五（节假日除外），每日交易时间为 9:30—11:30 和 13:00—15:00。在股票交易的这段时间内，股价的价格、涨跌幅、成交量等一系列数据是随时间变化的。对于一些想利用股票数据进行研究的学者，他们可能想要采集的是自开盘后，每隔十分钟或每隔一小时的股票的数据。针对这种定时采集任务，可以使用 Java 版的定时调度器 Quartz。

首先，在 com.job 包中编写一个实现 org.quartz.Job 接口的具体类 CarStockJob；同时，在该类的 execute()方法中编写具体所要实现的任务，如程序 7-26 所示。

程序 7-26

```java
package com.job;
import java.util.ArrayList;
import java.util.List;
import org.quartz.Job;
import org.quartz.JobExecutionContext;
import org.quartz.JobExecutionException;
import com.db.MYSQLControl;
import com.model.CarStockModel;
import com.parse.Parse;
import com.util.HttpRequestUtil;
public class CarStockJob implements Job {
    static HttpRequestUtil httpRequest = new HttpRequestUtil();
    public void execute(JobExecutionContext arg0) throws
```

```
JobExecutionException {
    for (int i = 1; i < 9; i++) {
        //拼接URL
        String url = "http://nufm.*****.com/EM_Finance2014"
                + "NumericApplication/JS.aspx?type=CT&token="
                + "4f1862fc3b5e77c150a2b985b12db0fd&sty="
                + "FCOIATC&js=(%7Bdata%3A%5B(x)%5D%2Crecords"
                + "Filtered%3A(tot)%7D)&cmd="
                + "C.BK04811&st=(ChangePercent)&sr=-1&p="
                + "" + i + "&ps=20&_=1551750725008";
        //请求拼接后的URL
        String content = httpRequest.getContentByHttpGetMethod(url,"utf-8");
        //解析URL
        List<CarStockModel> carstocks = new ArrayList<CarStockModel>();
        try {
            carstocks = Parse.getData(content);
        } catch (Exception e) {
            e.printStackTrace();
        }
        //存储数据
        MYSQLControl.insertCarStocks(carstocks);
    }
}
```

接着，在 com.main 中编写主类 CrawlerJobMain，如程序 7-27 所示。在程序 7-27 中设置的是每周一到周五的 9 点、10 点、11 点、13 点、14 点和 15 点分别执行一次采集任务。

程序 7-27

```
package com.main;
import static org.quartz.CronScheduleBuilder.cronSchedule;
import static org.quartz.JobBuilder.newJob;
import static org.quartz.TriggerBuilder.newTrigger;
import java.text.SimpleDateFormat;
import java.util.Date;
import org.quartz.CronTrigger;
import org.quartz.JobDetail;
```

```java
import org.quartz.Scheduler;
import org.quartz.SchedulerFactory;
import org.quartz.impl.StdSchedulerFactory;
import com.job.CarStockJob;
public class CrawlerJobMain {
    public void run() throws Exception {
        // 实例化任务调度器Scheduler
        SchedulerFactory sf = new StdSchedulerFactory();
        Scheduler sched = sf.getScheduler();
        //描述job实现类及其他相关的静态信息
        JobDetail job = newJob(CarStockJob.class).
                withIdentity("crawlerJob", "g").build();
        //每周一到周五的9点、10点、11点、13点、14点及15点分别执行一次程序
        CronTrigger trigger = newTrigger().withIdentity("crawlerTrigger", "g").
                withSchedule(cronSchedule("0 0 9,10,11,13,14,15 ? * MON-FRI")).build();
        Date ft = sched.scheduleJob(job, trigger);
        SimpleDateFormat sdf = new SimpleDateFormat("yyyy-MM-dd HH:mm:ss SSS");
        System.out.println(job.getKey() + " 已被安排执行于: "
                + sdf.format(ft) + ",并且以如下重复规则重复执行: " +
trigger.getCronExpression());
        //启动Scheduler
        sched.start();
    }
    public static void main(String[] args) throws Exception {
        CrawlerJobMain crawler = new CrawlerJobMain();
        crawler.run();
    }
}
```

在程序 7-27 中，cronSchedule(String cronExpression)方法中输入的是 Cron 表达式。由 org.quartz 中的 CronExpression 类可知，Cron 表达式共支持 7 种时间域（TimeZone），其使用语法为

秒 分 时 日 月 周 年(可选)

表 7.1 给出了 Cron 表达式支持的 7 种时间域的值范围和允许使用的特殊字符。其中，月和周使用英文格式时，不区分大小写，表 7-2 列举了各特殊字符的含义。

表 7.1 Quartz Cron 表达式支持的 7 种时间域

字 段 名	值 范 围	允许使用的特殊字符
秒	0～59	, - * /
分	0～59	, - * /
时	0～23	, - * /
日	1～31	, - * ? / L W
月	1～12 或 JAN—DEC	, - * /
周	1～7 或 SUN—SAT	, - * ? / L #
年（可选）	empty，1970—2099	, - * /

表 7.2 Cron 表达式中特殊字符的含义

特殊字符	含 义	案 例
,	指定多个并列的值	值 9,10,11 在小时域上表示 9 点、10 点、11 点
-	指定值的范围	在小时域上使用值 3-5，表示 3 点、4 点及 5 点
*	值域上的所有合法值	在小时域上使用，表示每小时都会触发
/	时间的递增	在小时域上使用 0/3，表示从 0 点开始每 3 小时触发一次
L	用在日上，表示一个月的最后一天；用在周上，表示一个月的最后一个星期	用在 2019 年 3 月份，表示 31 日触发
W	仅能用于日域中	1W 表示离每月 1 日最近的工作日
#	表示每个月第几个星期几	4#2 表示某月的第二个星期三

在实际应用中，可以利用 Cron 表达式在线工具自动生成符合需求的表达式，如 http://www.*****.net/，其通过单击的方式便可以生成 Cron 表达式，如图 7-25 所示。

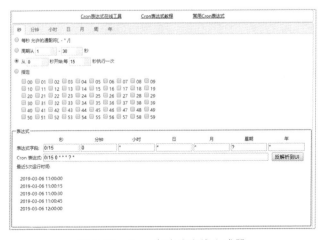

图 7-25 Cron 表达式在线生成器

7.4 本章小结

本章以三个网络爬虫实战案例为基础，重点介绍了一个开发网络爬虫的框架。框架主要包含以下 package，即 com.（db、main、model、parse、util）。其中，com.db 主要负责操作数据库和数据表，com.model 主要负责存放实体类，com.util 负责请求 URL 获取网页内容、处理文本、处理时间等，com.parse 主要负责网页内容的解析，com.main 主要负责存放相关主程序。在实际应用过程中，开发者只需要根据自身需求（例如模拟登录、定时执行任务等）扩展该框架的内容，便可以轻松实现数据的采集。

第 8 章

Selenium 的使用

8.1 Selenium 简介

Selenium 最初是由 Shinya Kasatani 基于火狐（Firefox）浏览器开发的工具，其主要用于网站的自动化测试。读者可以在火狐浏览器中安装 Selenium IDE 插件，并使用该插件录制在浏览器中的执行动作（如表单提交、单击和鼠标的移动等）。在本章中，我们将重点介绍 Selenium WebDriver 的使用。Selenium WebDriver 主要应用于程序（Java、Python 和 C#等）与浏览器的交互，其可以用来实现数据的采集。

Selenium 不自带浏览器，需要与第三方浏览器结合使用，如本章与 Selenium 结合使用的浏览器为火狐浏览器（版本号为 56.0 64 位）。相比 Jsoup 和 HttpClient 等工具，Selenium 有其特有优势，如自动加载页面（执行 JavaScript 脚本）、模拟真实的浏览器操作（可用于模拟登录）等。

8.2 Java Selenium 环境搭建

首先，创建一个 Java Maven 工程，并在该工程下添加两个文件夹 drivers 和 libs，如图 8.1 所示。

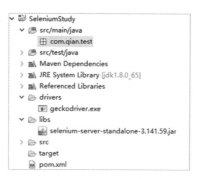

图 8.1 Java Maven 工程的项目结构

其次，在 Java 中使用 Selenium，需要下载相关 jar 包，在相关界面（见图 8.2）下载 selenium-server-standalone-3.141.59.jar。下载完成后，将该 jar 包放到工程的 libs 目录下，并将该 jar 包引入项目。另外一种配置相关 jar 包的方式是单击图 8.3 中所示的 "Download" 按钮，下载 selenium-java-3.141.59 压缩包，并进行解压，之后将解压后文件夹中的所有 jar 包放到 libs 目录下，并将这些 jar 包引入项目。再者，也可以直接使用 Maven 工程中的 pom.xml 文件配置所需的 jar 包。

```xml
<!-- https://*************.com/artifact/org.seleniumhq.selenium/selenium-java -->
    <dependency>
        <groupId>org.seleniumhq.selenium</groupId>
        <artifactId>selenium-java</artifactId>
        <version>3.141.59</version>
    </dependency>
```

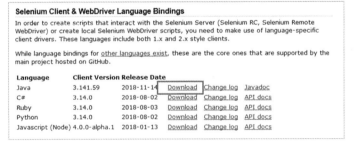

图 8.2　selenium-server-standalone-3.141.59 压缩包的下载

图 8.3　selenium-java-3.141.59 压缩包的下载

在相关 jar 包配置完成之后，需要下载 Selenium WebDriver 连接火狐浏览器的工具 geckodriver。

此处使用的火狐浏览器的版本号为 56，因此这里下载 geckodriver-v0.18.0-win64.zip，如图 8.4 所示。下载完成后，解压该压缩文件，并将文件夹中的 geckodriver.exe 放置到项目的 drivers 目录下。

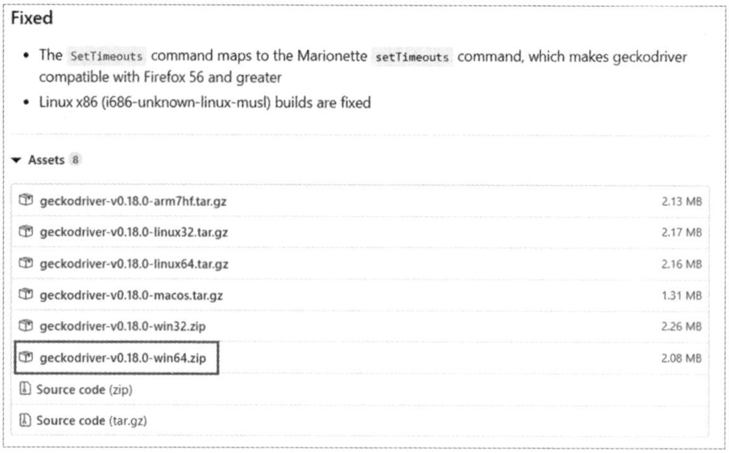

图 8.4 geckodriver-v0.18.0-win64.zip 的下载

8.3 浏览器的操控

搭建好 Java Selenium 环境后，便可以利用 Java 程序操控火狐浏览器，获取网页的响应内容。下面以某搜索网站为例，主要任务是使用 Selenium WebDriver 和 geckodriver 启动火狐浏览器，打开某搜索网站首页，并在搜索框中自动输入"Java 网络爬虫"，执行搜索，最后针对响应页面使用 Xpath 语法解析得到搜索结果的标题和 URL。

程序 8-1 为该案例的详细代码。这段代码中使用 setProperty(String key, String value) 配置 geckodriver，然后实例化 FirefoxDriver（声明使用的是火狐浏览器），接着使用火狐浏览器打开页面执行一系列操作。值得注意的是在执行搜索的过程中，需要利用 implicitlyWait() 方法休息一定的时间，以供网页完全加载数据。最后，使用 Xpath 语法进行元素定位，解析相应的字段。

程序 8-1

```java
package com.qian.test;
import java.util.List;
import java.util.concurrent.TimeUnit;
import org.openqa.selenium.By;
import org.openqa.selenium.WebElement;
import org.openqa.selenium.firefox.FirefoxDriver;
public class Test {
  public static void main(String[] args) {
    //geckodriver配置
```

```java
        System.setProperty("webdriver.gecko.driver", "drivers\\geckodriver.exe");
        //声明使用的是火狐浏览器
        FirefoxDriver driver = new FirefoxDriver();
        //使用火狐浏览器打开某搜索网站首页
        driver.get("http://www.*****.com");
        //元素定位,输入内容
        driver.findElementById("kw").sendKeys("Java 网络爬虫");
         //元素定位,执行单击命令
        driver.findElementById("su").click();
        //给出一定的响应时间
        driver.manage().timeouts().implicitlyWait(5, TimeUnit.SECONDS);
        driver.getPageSource();
        //使用Xpath解析数据
        List<WebElement> titleList =
                driver.findElements(By.xpath("//*[@class='result c-container ']/h3/a"));
        //输出新闻标题
        for(WebElement e : titleList){
            System.out.println("标题为: " + e.getText() + "\t" + "url为:"
                    + e.getAttribute("href"));
        }
        driver.quit();   // 关闭浏览器
    }
}
```

执行程序 8-1,会发现相关操作都会在浏览器中自动执行。另外,在控制台也会输出解析的标题和 URL,如图 8.5 所示。

图 8.5　控制台输出的内容

8.4 元素定位

在使用 Selenium 时，往往需要先通过定位器找到相应的元素，然后再进行其他操作。例如，在程序 8-1 中，先使用 findElementById("kw") 定位到搜索框，随后再利用 sendKeys(CharSequence... keysToSend) 方法输入搜索内容。定位器是一种抽象查询语言，功能是定位元素。Selenium WebDriver 提供了多种定位策略，如 id 定位、name 定位、class 定位、tag name 定位、link text 定位、Xpath 定位和 CSS 定位等。下面将分别介绍这些定位策略。

8.4.1 id 定位

id 定位，即通过 id 在网页中查询元素，使用简单且常用。例如，我们想要通过 id 定位百度首页的搜索框。首先，在谷歌浏览器中打开该页面，在搜索框的位置右击"检查"，便会看到图 8.6 所示的内容，进而可以确定搜索框可以通过 id='kw' 进行定位。同理，我们可以通过 id='su' 定位开始搜索按钮，如图 8.7 所示。

图 8.6　定位搜索框

图 8.7　定位按钮

在 Java 中，有以下四种方式实现 id 定位策略，如程序 8-2 所示。

程序 8-2

```
//第一种方式返回WebElement
driver.findElementById(<element id>)
//第二种方式返回List<WebElement>
driver.findElementsById(<element id>)
//第三种方式返回WebElement
driver.findElement(By.id(<element id>))
//第四种方式返回List<WebElement>
driver.findElements(By.id(<element id>))
```

8.4.2　name 定位

当元素中包含 name 属性时，可以使用 name 进行定位。如图 8.6 中的搜索框也可以使用 name='wd' 定位。在 Java 中，有以下四种方式实现 name 定位策略，如程序 8-3 所示。

程序 8-3

```
//第一种方式返回WebElement
driver.findElementByName (<element name>)
//第二种方式返回List<WebElement>
driver.findElementsByName (<element name>)
//第三种方式返回WebElement
driver.findElement(By.Name(<element name>))
//第四种方式返回List<WebElement>
driver.findElements(By.Name(<element name>))
```

8.4.3 class 定位

当元素中包含 class 属性时，可以使用 class 进行定位。如图 8.7 中的按钮也可以使用 class='btn self-btn bg s_btn' 定位。在 Java 中，有以下四种方式实现 class 定位策略，如程序 8-4 所示。

程序 8-4

```
//第一种方式返回 WebElement
driver.findElementByClassName (<element classname>)
//第二种方式返回 List<WebElement>
driver.findElementsByClassName(<element classname>)
//第三种方式返回 WebElement
driver.findElement(By.ClassName (<element classname>))
//第四种方式返回 List<WebElement>
driver.findElements(By.ClassName (<element classname>))
```

8.4.4 tag name 定位

Selenium WebDriver 也提供了 tag name（标签名称）定位的方法，使用标签名称可以很方便地定位一些元素，如定位所有表格中的<tr>、定位所有链接<a>等。程序 8-5 展示了实现 tag name 定位策略的四种方式。

程序 8-5

```
//第一种方式返回 WebElement
driver.findElementByTagName (<tagName>)
//第二种方式返回 List<WebElement>
driver.findElementsByTagName(<tagName>)
//第三种方式返回 WebElement
driver.findElement(By.tagName (<tagName>))
//第四种方式返回 List<WebElement>
driver.findElements(By.tagName(<element tagName>))
```

8.4.5 link text 定位

link text 定位，是通过链接文本定位链接的。在 Java 中，有以下四种方式实现 link text 定位策略，如程序 8-6 所示。

程序 8-6

```
//第一种方式返回 WebElement
driver.findElementByLinkText(<linkText>)
//第二种方式返回 List<WebElement>
driver.findElementsByLinkText(<linkText>)
//第三种方式返回 WebElement
driver.findElement(By.linkText(<linkText>))
//第四种方式返回 List<WebElement>
driver.findElements(By.linkText(<linkText>))
```

8.4.6 Xpath 定位

Selenium WebDriver 支持使用 Xpath 表达式定位元素，如程序 8-1 中使用 Xpath 表达式定位一些 <a> 元素。Xpath 使用路径表达式来定位 HTML 或 XML 文档中的节点或节点集合，在 5.1.2 节中已详细介绍了 Xpath 语法，这里不再赘述。在 Java 中，同样可以使用四种方式实现 Xpath 定位策略，如程序 8-7 所示。

程序 8-7

```
//第一种方式返回 WebElement
driver.findElementByXPath(<Xpath query expression>)
//第二种方式返回 List<WebElement>
driver.findElementByXPath(<Xpath query expression>)
//第三种方式返回 WebElement
driver.findElement(By.xpath(<Xpath query expression>))
//第四种方式返回 List<WebElement>
driver.findElements(By.xpath(<Xpath query expression>))
```

8.4.7 CSS 选择器定位

在 5.1.1 节中，介绍了 CSS 选择器，这里不再赘述。Selenium WebDriver 在定位元素时，也可以使用 CSS 选择器，其实现方式有四种，如程序 8-8 所示。

程序 8-8

```
//第一种方式返回 WebElement
driver.findElementByCssSelector(<css selector>)
//第二种方式返回 List<WebElement>
driver.findElementsByCssSelector(<css selector>)
//第三种方式返回 WebElement
```

```
driver.findElement(By.cssSelector(<css selector>))
//第四种方式返回List<WebElement>
driver.findElements(By.cssSelector(<css selector>))
```

8.5 模拟登录

Selenium 可以精准地定位浏览器中的元素（如输入框）、模拟真实的浏览器操作（如输入文本、单击等）。针对一些复杂且需要登录才能获取数据的网站，可以利用 Selenium 的特性，模拟登录该网站，获取登录的 Cookie。之后，使用 Jsoup 或者 Httpclient 采集该网站中的数据。以下，将以某网站的模拟登录为例，介绍 Selenium 的使用情况。

首先，使用谷歌或火狐浏览器，打开该网站登录页面。之后，在用户名、密码和登录按钮的位置分别右击"检查"，获取这三个元素的定位信息。由图 8.8 可知，可以使用 name='email'定位用户名框。同理，可以通过 id='password'定位密码框，通过 id='login'定位登录按钮。

图 8.8　定位用户名对应的表单

利用 Selenium，输入用户名和密码，并单击"登录"按钮执行登录操作。在登录完成后，使用 Jsoup 请求个人页面的数据，如图 8.9 所示。

图 8.9　登录后需要采集的数据

程序 8-9 给出了为模拟登录的完整代码。需要注意的是在模拟单击登录按钮后，需要使用 sleep(long millis)方法休息一段时间，使登录信息加载完整。在 Jsoup 请求指定页面时，使用了登录后的 Cookie 信息。执行程序 8-9，会发现相关登录操作会在浏览器中自动执行。同时，控制台也会输出解析得到的内容，如图 8.10 所示。

程序 8-9

```java
package com.qian.test;
import java.io.IOException;
import java.util.Set;
import java.util.concurrent.TimeUnit;
import org.jsoup.Connection.Response;
import org.jsoup.Jsoup;
import org.jsoup.nodes.Document;
import org.jsoup.nodes.Element;
import org.openqa.selenium.Cookie;
import org.openqa.selenium.firefox.FirefoxDriver;

public class LoginRenren {
    public static void main(String[] args) throws IOException, InterruptedException {
        //geckodriver配置
        System.setProperty("webdriver.gecko.driver", "drivers\\geckodriver.exe");
        //声明使用的是火狐浏览器
        FirefoxDriver driver = new FirefoxDriver();
        //使用火狐浏览器打开该网站
        driver.get("http://sns.******.com/");
        //元素定位，提交用户名及密码
        driver.findElementByName("email").clear();  //清空后输入
        driver.findElementByName("email").sendKeys("你的用户名");
        driver.manage().timeouts().implicitlyWait(5, TimeUnit.SECONDS);
        driver.findElementById("password").clear(); //清空后输入
        driver.findElementById("password").sendKeys("你的密码");
        //元素定位，单击登录按钮
        driver.findElementById("login").click();
        //休息一段时间，使得网页充分加载。注意这里非常有必要
        Thread.sleep(10*1000);
        Set<Cookie> cookies = driver.manage().getCookies();
        //获取登录的cookies
        String cookieStr = "";
        for (Cookie cookie : cookies) {
```

```java
                    cookieStr += cookie.getName() + "=" + cookie.getValue()
+ "; ";
            }
            System.out.println(cookieStr);
            //基于Jsoup，使用cookies请求个人信息页面
            Response orderResp = Jsoup    //添加一些header信息
                    .connect
("http://www.******.com/427727657/profile?v=info_timeline")
                    .header("Host", "www.******.com")
                    .header("Connection", "keep-alive")
                    .header("Cache-Control", "max-age=0")
                    .header("Accept", "text/html,application/xhtml+xml,
application/xml;q=0.9,image/webp,*;q=0.8")
                    .header("Origin", "http://www.******.com")
                    .header("Referer", "http://www.******.com/SysHome.do")
                    .userAgent("Mozilla/5.0 (Windows NT 10.0; Win64; x64;
rv:56.0) Gecko/20100101 Firefox/56.0")
                    .header("Content-Type", "application/
x-www-form-urlencoded")
                    .header("Accept-Encoding", "gzip, deflate, br")
                    .header("Upgrade-Insecure-Requests", "1")
                    .cookie("Cookie", cookieStr)
                    .execute();
            //解析数据
            Document doc = orderResp.parse();
            //System.out.println(doc);
            org.jsoup.select.Elements elements = doc.select
("div[class=info-section-info]")
                    .select("dl[class=info]");
            for (Element element : elements) {
                if (element.select("dt").text().contains("大学")) {
                    System.out.println(element.text());
                }
            }
            driver.quit();   // 关闭浏览器
        }
    }
```

第 8 章　Selenium 的使用　283

```
1555294974294    Marionette    DEBUG    Received DOM event "DOMContentLoaded" for "http://sns.    .com/
1555294975051    Marionette    DEBUG    Received DOM event "pageshow" for "http://sns    .com/"
1555294975729    Marionette    DEBUG    Canceled page load listener because no navigation has been detect
_de=EF0AC1051DCEE230196E82EB7F3536DEBB8C2103DE356; ver=7.0; _r01_=1; ln_hurl=http://hdn.    .cn/photo
大学    - 2011年- 管理学院    - 2011年- 管理学院
```

图 8.10　控制台输出的内容

8.6　动态加载 JavaScript 数据（操作滚动条）

在使用 Jsoup 和 Httpclient 直接请求 URL 时，有时会发现响应得到的 HTML 中包含的信息不全，未展示出的信息必须通过执行页面中的 JavaScript 代码才能展示。相比于 Jsoup 和 Httpclient，Selenium 可以利用 JavascriptExecutor 接口执行任意 JavaScript 代码。下面将以国外某网站中的新闻数据采集为例，讲解如何使用 Selenium 动态加载 JavaScript 数据。

在进入新闻的某一网页时，下拉滚动条，会发现底部展示了一系列的相关新闻（网站推荐的内容），如图 8.11 所示，继续下拉滚动条，新的新闻内容会不断地加载出来。

图 8.11　下拉滚动条所显示的其他相关新闻

为了采集每条新闻对应的一系列相关新闻，则需要不断地向下加载数据，即下拉浏览器滚动条。程序 8-10 给出了动态加载 JavaScript 数据的代码。在程序 8-10 中，使用 JavascriptExecutor 接口的 executeScript(String script, Object... args)方法执行

JavaScript 代码，具体的 JavaScript 代码为 scrollTo 方法，即页面滚动方法。在执行每一次滚动时，都需要休息一定的时间，以使得网页充分加载数据。在滚动完成之后，获取该网页的 HTML 内容并使用 Jsoup 工具解析相关新闻数据。执行程序 8-10，会在控制台输出新闻的标题和 URL，如图 8.12 所示。

程序 8-10

```java
package com.qian.test;
import java.io.IOException;
import org.jsoup.Jsoup;
import org.jsoup.nodes.Document;
import org.jsoup.nodes.Element;
import org.jsoup.select.Elements;
import org.openqa.selenium.JavascriptExecutor;
import org.openqa.selenium.firefox.FirefoxDriver;
public class FinancialNewsRolling {

    public static void main(String[] args) throws IOException, InterruptedException {
        //geckodriver配置
        System.setProperty("webdriver.gecko.driver", "drivers\\geckodriver.exe");
        //声明使用的是火狐浏览器
        FirefoxDriver driver = new FirefoxDriver();
        //使用火狐浏览器打开任意新闻页
        driver.get("https://www.***************.com/industry"
                + "/nclat-rejects-hdfc-plea-for-insolvency-"
                + "proceedings-against-rhc-holding/1640159/");
        // 执行JS操作
        JavascriptExecutor JS = (JavascriptExecutor) driver;
        Thread.sleep(3000);
        try {
            JS.executeScript("scrollTo(0, 5000)");
            System.out.println("1");
            Thread.sleep(5000);          //调整休眠时间可以获取更多的内容
            JS.executeScript("scrollTo(5000, 10000)");
            System.out.println("2");
            Thread.sleep(5000);
            JS.executeScript("scrollTo(10000, 30000)"); // 继续下拉
            System.out.println("3");
            Thread.sleep(5000);
            JS.executeScript("scrollTo(10000, 50000)"); //继续下拉
```

```java
            System.out.println("4");
        } catch (Exception e) {
            System.out.println("Error at loading the page ...");
            driver.quit();
        }
        String html = driver.getPageSource();
        //System.out.println(html);
        //解析数据
        Document doc = Jsoup.parse(html);
        Elements elements = doc.select("[id=taboola-below-article]")
                .select("div[id~=taboola-below-article-p?]");
        for (Element ele : elements) {
            String newsTitle = ele.select("a[class= item-label-href]")
                    .attr("title");
            String newsUrl = ele.select("a[class= item-label-href]")
                    .attr("href");
            System.out.println(newsTitle + "\t" + newsUrl);
        }
        driver.quit();   // 关闭浏览器
    }
}
```

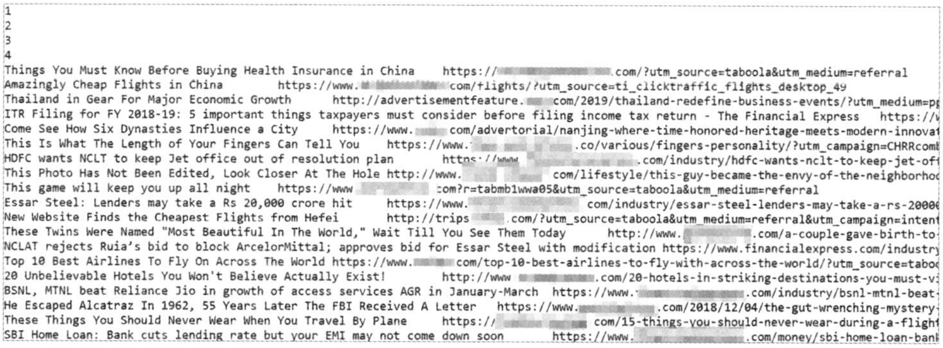

图 8.12　控制台输出的新闻标题和新闻 URL

8.7　隐藏浏览器

　　由以上内容可知使用 Selenium 采集网页数据时，需要不断地调用浏览器。实际上，通过对 Selenium 的设置，可以达到隐藏浏览器的效果。仍以 8.5 节中的新闻页面为例，程序 8-11 中给出了采集该新闻页面标题的代码。在程序 8-11 中，对火狐浏览器设置了 headless，其作用是实现无界面状态。同时，这里使用 while 循环的方式，以防止加载浏览器请求页面失败。最后，程序执行 JavaScript 代码，获取新闻的标题。执行程序 8-11，会在控制台输出新闻的标题信息。

程序 8-11

```java
package com.qian.test;
import java.io.IOException;
import java.util.concurrent.TimeUnit;
import org.openqa.selenium.JavascriptExecutor;
import org.openqa.selenium.firefox.FirefoxBinary;
import org.openqa.selenium.firefox.FirefoxDriver;
import org.openqa.selenium.firefox.FirefoxOptions;

public class JavaScriptTest {

    public static void main(String[] args) throws IOException, InterruptedException {
        FirefoxBinary firefoxBinary = new FirefoxBinary();
        firefoxBinary.addCommandLineOptions("--headless");
        //设置路径
        System.setProperty("webdriver.gecko.driver", "drivers\\geckodriver.exe");
        FirefoxOptions firefoxOptions = new FirefoxOptions();
        firefoxOptions.setBinary(firefoxBinary);
        FirefoxDriver driver = new FirefoxDriver(firefoxOptions);
        //直到加载该网页为止
        while (true){
            try{
                driver.get("https://www.****************.com"
                    + "/industry/nclat-rejects-hdfc-plea"
                    + "-for-insolvency-proceedings-against"
                    + "-rhc-holding/1640159/");
            }
            catch (Exception e)
            {
                driver.quit();
                driver = new FirefoxDriver(firefoxOptions);
                driver.manage().timeouts()
                    .pageLoadTimeout(10, TimeUnit.SECONDS);
                continue;
            }
            break;
```

```
        }
        //滚动条操作
        JavascriptExecutor JS = (JavascriptExecutor) driver;
        // 执行JS操作，返回新闻的标题
        String title = (String)JS.executeScript("return document.title");
        System.out.println(title);
        driver.quit();    // 关闭浏览器
    }
}
```

8.8 截取验证码

在网络爬虫中，很多网站会采用验证码的方式来反爬虫，例如在登录时设置验证码、频繁访问时自动弹出验证码等。针对模拟登录时的验证码输入问题，一种简单的解决方案是将验证码保存到本地，之后在程序中手动输入即可。但对于采集每页都需要验证码的网站来说，则需要使用验证码识别算法或调用一些 OCR API 来自动识别验证码，以保证其效率。

使用 Jsoup 或者 Httpclient，可以直接将验证码下载到本地。而对 Selenium 来说，可以使用截图的方式截取验证码，并保存到本地。下面，以某搜索网站为例介绍 Selenium 如何实现截取验证码，在程序中手动输入验证码内容，实现数据的采集。图 8.13 展示了需要输入的关键词及关键词对应的部分文章数据。

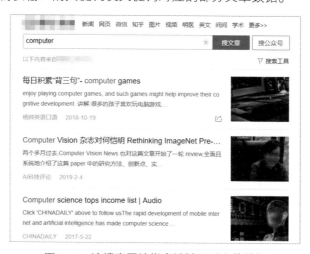

图 8.13   该搜索网站指定关键词对应的数据

在浏览器中打开网页时，会出现图 8.14 所示的情况，即访问异常，需要输入验证码。为此，我们使用 Selenium 截取验证码。首先，利用浏览器检查元素，使用 id='seccodeImage'定位到验证码，接着使用 getScreenshotAs 以及 getSubimage 方法截取该验证码。程序 8-12 给出了完整的数据采集代码。在手动输入验证码之后，单击"提交"按钮，便能看到所需内容。

图 8.14　访问异常界面

执行程序 8-12，会在"E:/钱洋个人/IdentifyingCode/"目录下出现 test.png 图片文件，如图 8.15 所示。在控制台输入验证码内容"8628bs"，便可以成功获取关键词对应的文章数据，如图 8.16 所示。

程序 8-12

```java
package com.qian.test;
import java.awt.image.BufferedImage;
import java.io.BufferedReader;
import java.io.File;
import java.io.IOException;
import java.io.InputStreamReader;
import java.util.concurrent.TimeUnit;
import javax.imageio.ImageIO;
import org.jsoup.Jsoup;
import org.jsoup.nodes.Document;
import org.jsoup.nodes.Element;
import org.jsoup.select.Elements;
import org.openqa.selenium.By;
import org.openqa.selenium.OutputType;
import org.openqa.selenium.Point;
import org.openqa.selenium.TakesScreenshot;
import org.openqa.selenium.WebElement;
import org.openqa.selenium.firefox.FirefoxBinary;
import org.openqa.selenium.firefox.FirefoxDriver;
import org.openqa.selenium.firefox.FirefoxOptions;
```

```java
    public class ScreenshotTest {

    public static void main(String[] args) throws IOException,
 InterruptedException {
            FirefoxBinary firefoxBinary = new FirefoxBinary();
            firefoxBinary.addCommandLineOptions("--headless");
            //设置路径
            System.setProperty("webdriver.gecko.driver", "drivers\\
 geckodriver.exe");
            FirefoxOptions firefoxOptions = new FirefoxOptions();
            firefoxOptions.setBinary(firefoxBinary);
            FirefoxDriver driver = new FirefoxDriver(firefoxOptions);
            //直到加载该网页为止
            while (true){
                try{
                    driver.get("http://weixin.*****.com/antispider/?"
                            + "from=%2fweixin%3Ftype%3d2%26query"
                            + "%3dcomputer+%26ie%3dutf8%26s_from%"
                            + "3dinput%26_sug_%3dy%26_sug_type_%3d");
                }
                catch (Exception e)
                {
                    driver.quit();
                    driver = new FirefoxDriver(firefoxOptions);
                    driver.manage().timeouts()
                    .pageLoadTimeout(10, TimeUnit.SECONDS);
                    continue;
                }
                break;
            }
            WebElement webEle = driver.findElement(By.id("seccodeImage"));
            // Get entire page screenshot
            java.io.File screenshot = ((TakesScreenshot)driver)
                    .getScreenshotAs(OutputType.FILE);
            BufferedImage fullImg = ImageIO.read(screenshot);
            Point point = webEle.getLocation();
            int eleWidth = webEle.getSize().getWidth();
            int eleHeight = webEle.getSize().getHeight();
            BufferedImage eleScreenshot = fullImg.getSubimage(point.
 getX(), point.getY(),
                    eleWidth, eleHeight);
            ImageIO.write(eleScreenshot, "png", new File("E:/钱洋个人
```

```
/IdentifyingCode/test.png"));
        System.out.println("请输入验证码：");
        BufferedReader buff=new BufferedReader(new InputStreamReader(System.in));
        String captcha_solution="";
        try {
            captcha_solution = buff.readLine();
        } catch (IOException e) {
            e.printStackTrace();
        }
        driver.findElement(By.name("c")).sendKeys(captcha_solution);
        driver.findElementById("submit").click();
        Thread.sleep(10*1000);    //休息一段时间，使得网页充分加载。这里非常有必要
        String html = driver.getPageSource();
        Document doc = Jsoup.parse(html);
        Elements elements = doc.select("div[class=txt-box]");
        for (Element ele : elements) {
            String newsTitle = ele.select("h3").select("a").text();
            String newsUrl = ele.select("h3").select("a").attr("href");
            System.out.println(newsTitle + "\t" + newsUrl);
        }
        driver.quit();  // 关闭浏览器
    }
}
```

图 8.15 验证码图片

图 8.16 控制台输出的内容

8.9 本章小结

本章主要介绍了 Selenium WebDriver 在网络爬虫中的应用。相比 Jsoup、Httpclient 和 URLConnection，Selenium 的主要优势是可以与浏览器进行交互（如输入文字、单击按钮等）及执行加载 JavaScript 代码。但 Selenium 也存在缺点，如每执行一个 URL 都相当于在浏览器中打开一个网页，并且需要加载网页中的 JavaScript 代码，因此，Selenium 的效率较低，仅适用于小规模数据采集。

第 9 章

网络爬虫开源框架

9.1 Crawler4j 的使用

9.1.1 Crawler4j 简介

Crawler4j 是由 Yasser Ganjisaffar 开发的一个简单易用的开源网络爬虫框架,它支持多线程和深度数据采集,并且内置 URL 过滤机制(由 frontier 包实现)。同时,针对 URL 对应的页面内容,开发者可利用数据解析工具(如 Jsoup)提取网页中的结构化字段。Crawler4j 项目的源码可以在 GitHub 上进行下载。

9.1.2 jar 包的下载

首先,在 Maven 工程的 pom.xml 文件中配置 Crawler4j 的相关 jar 包及相应的依赖 jar 包。

```xml
<!-- https://*************.com/artifact/edu.uci.ics/crawler4j -->
<dependency>
    <groupId>edu.uci.ics</groupId>
    <artifactId>crawler4j</artifactId>
    <version>4.4.0</version>
</dependency>
```

在 Crawler4j 项目中,共包含 7 个 package,即 edu.uci.ics.crawler4j.(crawler、fetcher、parser、robotstxt、url、util),如图 9.1 所示。每个 package 都有各自的功能。

edu.uci.ics.crawler4j.crawler 负责网络爬虫的配置和信息的采集,相关类有 CrawlConfig.java、WebCrawler.java、CrawlController.java 和 Page.java 等。其中,WebCrawler.java 类通过实现 Runnable 接口来实现多线程数据采集。

edu.uci.ics.crawler4j.fetcher 负责具体页面的请求与处理,相关类主要是在 Httpclient 包的基础上开发的,如 PageFetcher.java 和 PageFetchResult.java。

edu.uci.ics.crawler4j.frontier 负责 URL 集合的调度，如为 URL 分配 docid（即 URL 编号）、URL 过滤、URL 添加到集合等。

edu.uci.ics.crawler4j.parser 负责解析存储在 Page 对象里的内容，解析的数据类型包括二进制数据、文本数据和 HTML 数据。

edu.uci.ics.crawler4j.robotstxt 负责配置 robots 协议（爬虫协议）、确定协议是否存在、根据协议判断 URL 信息是否被允许采集。

edu.uci.ics.crawler4j.url 负责规范 URL、URL 路径处理、URL 信息封装等。

edu.uci.ics.crawler4j.util 为基础工具类。

图 9.1　Crawler4j 项目中的 packages

9.1.3　入门案例

本节将利用 Crawler4j 构建一个简单的数据采集程序。采集的网站是某资讯类平台。

打开该网站的 URL，发现网页中存在文本类新闻、图片类新闻和视频类新闻等内容，如图 9.2 所示。而本节关注的是文本类新闻，其 URL 的类型需要以"https://www.******.com"为前缀，以".htm"为后缀，如 https://www.******.com/movies/special/priyanka-deepika-defeat-salman-khan/20190711.htm。

使用 Crawler4j 构建一个网络爬虫，分为两个步骤。

步骤 1：创建一个爬虫类，该类需要继承 WebCrawler，具体如程序 9-1 所示。从程序 9-1 中可以看出 RediffCrawler 重写了 WebCrawler 类的 shouldVisit()方法和 visit()方法。其中，shouldVisit()方法主要用于设计页面访问规则，如这里访问的页面 URL 必须以"https://www.******.com/"为前缀，以".htm"为后缀。visit()方法主要用于处理页面结果，如本程序中针对已访问的 URL 页面，将新闻 id（程序自动编号）、解析的新闻标题和新闻的 HTML 内容写入指定目录的文件中。

程序 9-1

```java
package com.crawler.rediff;

import edu.uci.ics.crawler4j.crawler.WebCrawler;
import edu.uci.ics.crawler4j.parser.HtmlParseData;
import java.io.FileWriter;
import java.io.IOException;
import java.util.regex.Pattern;
import edu.uci.ics.crawler4j.crawler.Page;
import edu.uci.ics.crawler4j.url.WebURL;
public class RediffCrawler extends WebCrawler {
    //设置正则规则以".htm"为后缀
    private final static Pattern URLPattern = Pattern.compile(".*(\\.htm)$");
    /**
     * 用于过滤URL，如这里必须是以"https://www.******.com/"开头
     * 以".tml"结尾的URL才可以被访问
     */
    @Override
    public boolean shouldVisit(Page referringPage, WebURL url) {
        String href = url.getURL().toLowerCase();
        return URLPattern.matcher(href).matches() &&
                href.startsWith("https://www.******.com/");
    }

    /**
     * 处理URL
     */
    @Override
    public void visit(Page page) {
        String url = page.getWebURL().getURL(); // 获取URL
        if (URLPattern.matcher(url).matches()
                && page.getParseData() instanceof HtmlParseData) {
            FileWriter writer = null;
            try {
                //data目录下面存储每篇文档的内容
                writer = new FileWriter("data/"
                        +page.getWebURL().getDocid() + ".txt");
            } catch (IOException e) {
                e.printStackTrace();
            }
```

```
            HtmlParseData htmlParseData = (HtmlParseData) page
                    .getParseData();
            String html = htmlParseData.getHtml();
            try {
                writer.append("新闻的id为:"
                        + page.getWebURL().getDocid()
                        + "\n链接为: " + page.getWebURL().getURL()
                        + "\n新闻的标题为:" + htmlParseData.getTitle()
                        +"\n");
                writer.append(html + "\n");
            } catch (IOException e) {
                e.printStackTrace();
            }
            try {
                writer.close();
            } catch (IOException e) {
                e.printStackTrace();
            }
        }
    }
}
```

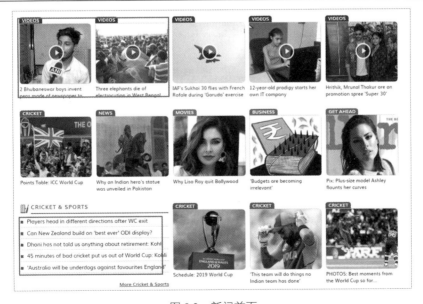

图 9.2 新闻首页

步骤 2：构建一个控制类，用于配置网络爬虫和执行数据采集任务，如程序 9-2 所示。在主方法中，设置了爬虫状态存储的文件夹、线程数目、采集深度、是否允许重定向和最多采集的页面数。执行主方法，可以看到控制台输出了一系列文件，同时在工程的"data/"目录下产生了 9 个 txt 文件，如图 9.3 所示。之所以产生 9 个文件而不是 10 个文件，是因为在程序 9-1 的 visit() 方法中添加了文件输出的判断条件，即必须以"数字"结尾的 URL。因此，编号为 1 的种子 URL 不符合条件。打开其中一个 txt 文件，如"2.txt"，可以看到新闻对应的内容信息，如图 9.4 所示。

程序 9-2

```java
package com.crawler.rediff;

import edu.uci.ics.crawler4j.crawler.CrawlConfig;
import edu.uci.ics.crawler4j.crawler.CrawlController;
import edu.uci.ics.crawler4j.fetcher.PageFetcher;
import edu.uci.ics.crawler4j.robotstxt.RobotstxtConfig;
import edu.uci.ics.crawler4j.robotstxt.RobotstxtServer;

public class RediffNewsController {

    public static void main(String[] args) throws Exception {
        //爬虫状态存储文件夹
        String crawlStorageFolder = "F:/program_work/java_work"
                + "/CSDNCourse/Crawler4j/data/craw/root";
        int numberOfCrawlers = 5;              //线程数
        CrawlConfig config = new CrawlConfig();
        config.setMaxDepthOfCrawling(1);       //只采集第一层页面的数据
        config.setMaxPagesToFetch(10);         //最多采集10个页面
        config.setFollowRedirects(false);      //是否允许重定向
        config.setCrawlStorageFolder(crawlStorageFolder);
        // 配置信息
        PageFetcher pageFetcher = new PageFetcher(config);
        //robots协议
        RobotstxtConfig robotstxtConfig = new RobotstxtConfig();
        RobotstxtServer robotstxtServer = new RobotstxtServer(robotstxtConfig, pageFetcher);
        CrawlController controller = new CrawlController(config,
                pageFetcher, robotstxtServer);
        //添加种子URL
        controller.addSeed("https://www.******.com");
        //运行网络爬虫
```

```
        controller.start(RediffCrawler.class, numberOfCrawlers);
    }
}
```

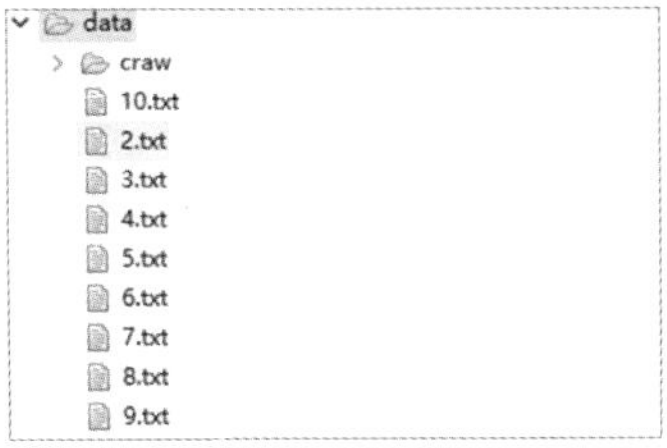

图 9.3　数据采集结果

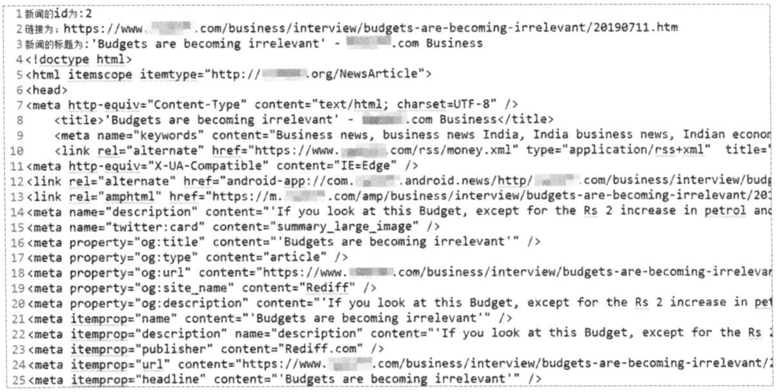

```
新闻的id为:2
链接为：https://www. .com/business/interview/budgets-are-becoming-irrelevant/20190711.htm
新闻的标题为:'Budgets are becoming irrelevant' - .com Business
<!doctype html>
<html itemscope itemtype="http:// .org/NewsArticle">
<head>
<meta http-equiv="Content-Type" content="text/html; charset=UTF-8" />
    <title>'Budgets are becoming irrelevant' - .com Business</title>
    <meta name="keywords" content="Business news, business news India, India business news, Indian econon
    <link rel="alternate" href="https://www. .com/rss/money.xml" type="application/rss+xml"  title="
<meta http-equiv="X-UA-Compatible" content="IE=Edge" />
<link rel="alternate" href="android-app://com. .android.news/http/ .com/business/interview/budg
<link rel="amphtml" href="https://m. .com/amp/business/interview/budgets-are-becoming-irrelevant/201
<meta name="description" content="'If you look at this Budget, except for the Rs 2 increase in petrol anc
<meta name="twitter:card" content="summary_large_image" />
<meta property="og:title" content="'Budgets are becoming irrelevant'" />
<meta property="og:type" content="article" />
<meta property="og:url" content="https://www. .com/business/interview/budgets-are-becoming-irrelevar
<meta property="og:site_name" content="Rediff" />
<meta property="og:description" content="'If you look at this Budget, except for the Rs 2 increase in pet
<meta itemprop="name" content="'Budgets are becoming irrelevant'" />
<meta itemprop="description" name="description" content="'If you look at this Budget, except for the Rs 2
<meta itemprop="publisher" content="Rediff.com" />
<meta itemprop="url" content="https://www. .com/business/interview/budgets-are-becoming-irrelevant/2
<meta itemprop="headline" content="'Budgets are becoming irrelevant'" />
```

图 9.4　“2.txt”中存储的数据

9.1.4　相关配置

在程序 9-2 中，使用 CrawlConfig 类中的方法配置了网络爬虫采集页面深度、采集页面数、是否允许重定向和状态文件存储位置方面的信息。查看 CrawlConfig 类的源码可以发现，其还可以配置网络爬虫的其他信息。

1. User-Agent

使用 CrawlConfig 类中的 setUserAgentString()方法可以配置 User-Agent，配置程序如下所示。

```
        config.setUserAgentString("Mozilla/5.0 (Windows NT 10.0; Win64; x64) "
                + "AppleWebKit/537.36 (KHTML, like Gecko) Chrome/63.0.3239.108 Safari/537.36");
```

2. 头信息

CrawlConfig 类中的 setDefaultHeaders()方法用于设置请求头信息，该方法的源码如程序 9-3 所示。从该方法的源码中可以发现，调用该方法需要输入 Collection<BasicHeader>类型的参数。为此，可以使用程序 9-4 的方式设置一系列的请求头信息。

程序 9-3

```
    /**
     * Set the default header collection (creating copies of the
provided headers).用于设置默认头
     */
    public void setDefaultHeaders(Collection<? extends Header> defaultHeaders) {
        Collection<BasicHeader> copiedHeaders = new HashSet<>();
        for (Header header : defaultHeaders) {
            copiedHeaders.add(new BasicHeader(header.getName(), header.getValue()));
        }
        this.defaultHeaders = copiedHeaders;
    }
```

程序 9-4

```
    HashSet<BasicHeader> collections = new HashSet<BasicHeader>();
    collections.add(new         BasicHeader("User-Agent","Mozilla/5.0 (Windows NT 10.0; Win64; x64) "
            + "AppleWebKit/537.36 (KHTML, like Gecko) Chrome/63.0.3239.108 Safari/537.36"));
    collections.add(new BasicHeader("Accept","text/html,application/xhtml+xml,application/xml;"
            + "q=0.9,image/webp,image/apng,*/*;q=0.8"));
    collections.add(new BasicHeader("Accept-Encoding", "gzip, deflate"));
    collections.add(new BasicHeader("Accept-Language", "zh-CN,zh;q=0.9"));
    collections.add(new BasicHeader("Content-Type","application/x-www-form-urlencoded;charset=UTF-8"));
    collections.add(new BasicHeader("Connection", "keep-alive"));
    config.setDefaultHeaders(collections);
```

3. 请求时间间隔

为了礼貌地采集网站数据，CrawlConfig 类提供了 setPolitenessDelay()方法，该方法用于设置每个线程任意两次请求之间的时间间隔（默认值为 200 毫秒）。以下为配置程序。

```
config.setPolitenessDelay(1000); //1000毫秒
```

4. HTTPS 网页的采集

针对使用 HTTPS 协议的 URL 页面内容的采集，CrawlConfig 类提供了 setIncludeHttpsPages()方法，该方法的默认输入值为 true。以下为配置程序。

```
config.setIncludeHttpsPages(false); //不采集HTTPS网页
```

5. 二进制文件

CrawlConfig 类提供了 setIncludeBinaryContentInCrawling()方法，用于设置是否采集图片和 PDF 等二进制文件，该方法的默认值为 false。以下为配置程序。

```
config.setIncludeBinaryContentInCrawling (true); //允许采集二进制文件
```

6. 超时时间

CrawlConfig 类中的 setSocketTimeout()方法用于设置获取数据的超时时间，默认值为 20000 毫秒；setConnectionTimeout()用于设置建立连接超时，默认值为 30000 毫秒。如下为这两种方法的使用示例。

```
config.setSocketTimeout(10000);
config.setConnectionTimeout(10000);
```

7. 代理使用

CrawlConfig 类提供了用于设置代理域名的方法——setProxyHost()方法及设置端口的方法——setProxyPort()方法。如下为这两种方法的使用示例。

```
config.setProxyHost("171.97.67.160");
config.setProxyPort(3128);
```

使用 setProxyHost()方法和 setProxyPort()方法只能配置单个代理，其并不支持代理的可用性检查和代理的切换。

9.1.5 图片的采集

9.1.3 节介绍了 HTML 类型数据的采集。下面，将通过一个案例介绍如何使用 Crawler4j 采集网页中的图片。采集的网站是 Pixabay。

在该网页的搜索框中，输入关键词，将会返回一些与该关键词有关的图片，如图 9.5 所示。本案例将提供"深林""汽车""动物""文字"四个关键词，并采集关键词相关的图片。

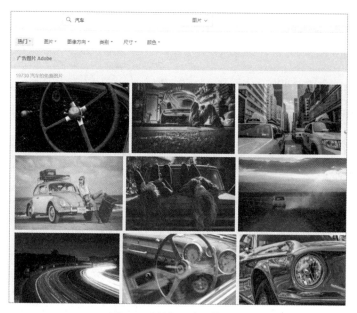

图 9.5 关键词对应的图片

根据 Crawler4j 的使用步骤，首先创建一个爬虫类，该类需要继承 WebCrawler，具体如程序 9-5 所示。在程序 9-5 中，定义正则表达式规则 FILTERS 用于过滤网页中的噪音 URL，定义 imgPatterns 来匹配图片 URL。其中，Configure()方法用于配置图片输出的目录。为了遍历到更多相关图片的 URL，在 shouldVisit()方法中设置了检测页面 URL 是否以"https://*******.com/zh/photos"为前缀。在 visit()方法中，设置了输出图片应满足的条件，如图片的格式必须以"bmp/gif"等为后缀、文件类型必须为二进制、图片的存储容量大于 5KB。

程序 9-5

```
package com.crawler.picture;

import edu.uci.ics.crawler4j.crawler.WebCrawler;
```

```java
import edu.uci.ics.crawler4j.parser.BinaryParseData;
import java.io.File;
import java.io.IOException;
import java.util.UUID;
import java.util.regex.Pattern;
import com.google.common.io.Files;
import edu.uci.ics.crawler4j.crawler.Page;
import edu.uci.ics.crawler4j.url.WebURL;
public class PictureCrawler extends WebCrawler {
    //设置过滤规则
    private final static Pattern FILTERS = Pattern
            .compile(".*(\\.(css|js|mid|mp2|mp3|mp4|wav|avi|mov|mpeg|ram|m4v|pdf" +
                    "|rm|smil|wmv|swf|wma|zip|rar|gz))$");
    /*
     * 匹配图片规则
     * JPG/JPEG/PNG格式
     */
    private static final Pattern imgPatterns = Pattern
            .compile(".*(\\.(bmp|gif|jpe?g|png|tiff?))$");
    private static File storageFolder; // 爬取的图片本地存储地址
    /**
     * 配置本地存储文件
     * @param storageFolderName
     */
    public static void configure(String storageFolderName) {
        storageFolder = new File(storageFolderName); //实例化
        if (!storageFolder.exists()) { // 假如文件不存在
            storageFolder.mkdirs();    // 创建一个文件
        }
    }

    @Override
    public boolean shouldVisit(Page referringPage, WebURL url) {
        String href = url.getURL().toLowerCase();
        if (FILTERS.matcher(href).matches()) {
            return false;
        }

        if (imgPatterns.matcher(href).matches()) {
            return true;
```

```java
            }
            if (href.startsWith("https://*******.com/zh/photos")) {
                return true;
            }
            return false;
        }
        /**
         * 处理URL，存储图片
         */
        @Override
        public void visit(Page page) {
            String url = page.getWebURL().getURL(); // 获取URL
            //满足条件输出图片
            if (imgPatterns.matcher(url).matches() &&
                    page.getParseData() instanceof BinaryParseData &&
                    page.getContentData().length > (5 * 1024)) {
                // 获取图片后缀
                String extension = url.substring(url.lastIndexOf('.'));
                // 通过UUID拼接成唯一图片名称
                String hashedName = UUID.randomUUID() + extension;
                // 存储图片
                String filename = storageFolder.getAbsolutePath() + "/"
 + hashedName;
                try {
                    // 将爬取到的文件存储到指定文件
                    Files.write(page.getContentData(), new
 File(filename));
                    System.out.println("stored url:" + url);
                } catch (IOException iox) {
                    iox.printStackTrace();
                }
            }
        }
    }
```

接着，构建一个控制类，用于配置网络爬虫、执行数据采集任务，如程序9-6所示。在主方法中，设置了爬虫状态存储的文件夹、线程数目、采集的深度、是否允许重定向、头信息、是否允许采集二进制文件和超时时间。执行程序9-6，可以看到成功下载了一系列与关键词相关的图片，如图9.6所示。

程序 9-6

```java
package com.crawler.picture;
import java.net.URLEncoder;
import java.util.HashSet;
import org.apache.http.message.BasicHeader;
import edu.uci.ics.crawler4j.crawler.CrawlConfig;
import edu.uci.ics.crawler4j.crawler.CrawlController;
import edu.uci.ics.crawler4j.fetcher.PageFetcher;
import edu.uci.ics.crawler4j.robotstxt.RobotstxtConfig;
import edu.uci.ics.crawler4j.robotstxt.RobotstxtServer;
public class PictureCrawlerController {
  public static void main(String[] args) throws Exception {
      //爬虫状态存储文件夹
      String crawlStorageFolder = "F:/program_work/java_work/CSDNCourse"
              + "/Crawler4j/data/craw/root";
      String storageFolder = "F:/picture/";
      int numberOfCrawlers = 5;    //线程数
      CrawlConfig config = new CrawlConfig();
      config.setMaxDepthOfCrawling(3);    //只采集第三层页面的数据
      config.setFollowRedirects(false);   //是否允许重定向
      config.setCrawlStorageFolder(crawlStorageFolder);
      //设置头信息
      HashSet<BasicHeader> collections = new HashSet<BasicHeader>();
      collections.add(new  BasicHeader("User-Agent","Mozilla/5.0 (Windows NT 10.0; "
              + "Win64; x64) AppleWebKit/537.36 (KHTML, like Gecko) "
              + "Chrome/63.0.3239.108 Safari/537.36"));
      collections.add(new BasicHeader("Accept","text/html, application/xhtml+xml,"
              + "application/xml;"
              + "q=0.9,image/webp,image/apng,*/*;q=0.8"));
      collections.add(new  BasicHeader("Accept-Encoding", "gzip, deflate"));
      collections.add(new BasicHeader("Accept-Language", "zh-CN,zh;q=0.9"));
      collections.add(new BasicHeader("Connection", "keep-alive"));
      config.setDefaultHeaders(collections);
      config.setPolitenessDelay(2000);        //礼貌采集
      //是否采集二进制文件
      config.setIncludeBinaryContentInCrawling(true);
      config.setSocketTimeout(10000);         //超时设置
```

```
        config.setConnectionTimeout(10000);        //超时设置
        // 配置信息
        PageFetcher pageFetcher = new PageFetcher(config);
        RobotstxtConfig robotstxtConfig = new RobotstxtConfig();
//robots协议
        RobotstxtServer robotstxtServer = new RobotstxtServer
(robotstxtConfig, pageFetcher);
        CrawlController controller = new CrawlController(config,
            pageFetcher, robotstxtServer);
        //添加种子URL
        controller.addSeed("https://*******.com/zh/images/search/"
            + URLEncoder.encode("深林", "utf-8") + "/");
        controller.addSeed("https://*******.com/zh/images/search/"
            + URLEncoder.encode("汽车", "utf-8") + "/");
        controller.addSeed("https://*******.com/zh/images/search/"
            + URLEncoder.encode("动物", "utf-8") + "/");
        controller.addSeed("https://*******.com/zh/images/search/"
            + URLEncoder.encode("文字", "utf-8") + "/");
        PictureCrawler.configure(storageFolder); // 配置存储位置
        //运行网络爬虫
        controller.start(PictureCrawler.class, numberOfCrawlers);
    }
}
```

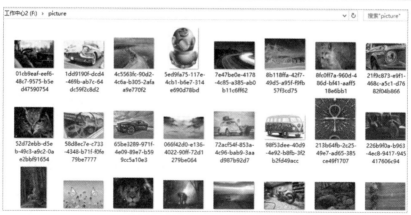

图 9.6 使用 Crawler4j 采集的图片

9.1.6 数据采集入库

本节将使用 Crawler4j 采集网页，并将网页内容解析成结构化数据存储到 MySQL 数据库。采集网站为 The Japan News，种子节点包括：

```
http://the-*****-news.com/news/business。        //商业新闻
http://the-*****-news.com/news/society。         //社会新闻
http://the-*****-news.com/news/sports。          //体育新闻
```

需要解析的字段为新闻 URL、新闻发表时间、新闻标题和新闻内容。

1. 框架内容

本案例使用第 7 章介绍的框架结构编写，框架中的 package 有 com.（db、main、model、parse）。图 9.7 所示为整个项目的结构。

在编写代码之前，使用 Maven 工程中的 pom.xml 文件配置整个项目需要的 jar 包。

```xml
<!-- https://*************.com/artifact/edu.uci.ics/crawler4j -->
<dependency>
    <groupId>edu.uci.ics</groupId>
    <artifactId>crawler4j</artifactId>
    <version>4.4.0</version>
</dependency>
<!-- https://*************.com/artifact/org.jsoup/jsoup -->
<dependency>
    <groupId>org.jsoup</groupId>
    <artifactId>jsoup</artifactId>
    <version>1.11.3</version>
</dependency>
<!-- 数据库相关 -->
<dependency>
    <groupId>mysql</groupId>
    <artifactId>mysql-connector-java</artifactId>
    <version>5.1.32</version>
</dependency>
<dependency>
    <groupId>commons-dbutils</groupId>
    <artifactId>commons-dbutils</artifactId>
    <version>1.7</version>
</dependency>
<dependency>
    <groupId>org.apache.commons</groupId>
    <artifactId>commons-dbcp2</artifactId>
    <version>2.5.0</version>
</dependency>
```

图 9.7　项目结构

2．程序设计

（1）com.Model

明确要采集的字段，即新闻 id、新闻 URL、新闻发表时间、新闻标题和新闻内容。为此，可构建 NewsModel 类用于封装采集的数据，如程序 9-7 所示。

程序 9-7

```
package com.model;
public class NewsModel {
 private int docid;          //Crawler4j自动编号的id
 private String url;         //新闻URL
 private String time;        //新闻发表时间
 private String title;       //新闻标题
 private String content;     //新闻内容
 public int getDocid() {
     return docid;
 }
 public void setDocid(int docid) {
     this.docid = docid;
 }
 public String getUrl() {
     return url;
 }
 public void setUrl(String url) {
     this.url = url;
 }
 public String getTime() {
     return time;
```

```
    }
    public void setTime(String time) {
        this.time = time;
    }
    public String getTitle() {
        return title;
    }
    public void setTitle(String title) {
        this.title = title;
    }
    public String getContent() {
        return content;
    }
    public void setContent(String content) {
        this.content = content;
    }
}
```

（2）com.parse

针对 Crawler4j 采集的 Page 类型的数据，设计数据解析程序。程序 9-8 所示为解析类 Parse，该程序中使用 Jsoup 解析 HTML 页面。在解析得到所有的字段之后，需要实例化 NewsModel，并使用 set()方法封装数据。

程序 9-8

```
package com.parse;
import org.jsoup.Jsoup;
import org.jsoup.nodes.Document;
import com.model.NewsModel;
import edu.uci.ics.crawler4j.crawler.Page;
import edu.uci.ics.crawler4j.parser.HtmlParseData;
public class Parse {
    /**
     * 解析新闻HTML数据
     *
     * @param Page
     * @return SinaAutoNews
     */
    public static NewsModel getData (Page page) {
        HtmlParseData htmlParseData = (HtmlParseData) page
                .getParseData();
        String html = htmlParseData.getHtml();
```

```java
        //使用Jsoup解析新闻网站
        Document doc = Jsoup.parse(html);
        //新闻id
        int docid = page.getWebURL().getDocid();
        //新闻URL
        String url = page.getWebURL().getURL();
        //新闻标题
        String title = doc.select("#articleContentHeaderH1")
                .select("span").text();
        //新闻时间
        String time = doc.select("#articleContentHeaderTime")
                .attr("datetime");
        //新闻内容
        String content = doc.select("#articleContentBody").text();
        //封装数据
        NewsModel model = new NewsModel();
        model.setDocid(docid);
        model.setUrl(url);
        model.setTitle(title);
        model.setTime(time);
        model.setContent(content);
        return model;
    }

}
```

（3）com.db

com.db 中包含两个 Java 文件，即 MyDataSource 与 MYSQLControl。MyDataSource 类的写法与第 7 章该类的写法一致。为存储采集的汽车新闻数据，需要在数据库中创建对应的数据表 japannews，以下为建表语句.

```sql
CREATE TABLE `japannews` (
  `docid` int(10) NOT NULL,
  `url` varchar(200) DEFAULT NULL,
  `time` varchar(100) DEFAULT NULL,
  `title` text,
  `content` longtext,
  PRIMARY KEY (`docid`)
) ENGINE=InnoDB DEFAULT CHARSET=utf8;
```

针对待存储的数据，需要编写数据库、数据表操作类 MYSQLControl，如程序 9-9 所示。程序的 executeInsert() 输入的数据类型为 NewsModel。

程序 9-9

```java
package com.db;

import java.sql.SQLException;
import javax.sql.DataSource;
import org.apache.commons.dbutils.QueryRunner;
import com.model._NewsModel;
public class MYSQLControl {
    //根据本地数据库地址修改
    static DataSource ds = MyDataSource.getDataSource
("jdbc:mysql://127.0.0.1:3306/"
            + "crawler?useUnicode=true&characterEncoding=UTF8");
    static QueryRunner qr = new QueryRunner(ds);
    public static void executeInsert(NewsModel model) {
        Object[][] params = new Object[1][5];   //数据的维度
        params[0][0] = model.getDocid();
        params[0][1] = model.getUrl();
        params[0][2] = model.getTime();
        params[0][3] = model.getTitle();
        params[0][4] = model.getContent();
        //使用batch方法插入
        try {
            qr.batch("insert into  japannews (docid,url,time,"
                    + "title,content) "
                    + "values (?,?,?,?,?)", params);
            System.out.println("执行数据库完毕! " + "成功插入数据 1条");
        } catch (SQLException e) {
            e.printStackTrace();
        }
    }
}
```

（4）com.main

在 com.main 中共包含两个类：JNewsCrawler 和 JNewsController。其中，JNewsCrawler 为爬虫类，其继承了 WebCrawler，具体如程序 9-10 所示。在程序 9-10 中，定义正则表达式规则 URLPattern 用于匹配符合条件的 URL，即以数字为后缀。在 shouldVisit() 方法中，添加了 URL 满足的前缀条件。在 visit() 方法中，先判断 Page 对应的数据是否为 HTML 类型，之后解析每个 Page，同时进行数据插入操作。

程序9-10

```java
package com.main;

import edu.uci.ics.crawler4j.crawler.WebCrawler;
import edu.uci.ics.crawler4j.parser.HtmlParseData;
import java.util.regex.Pattern;
import com.db.MYSQLControl;
import com.parse.Parse;
import edu.uci.ics.crawler4j.crawler.Page;
import edu.uci.ics.crawler4j.url.WebURL;
public class JNewsCrawler extends WebCrawler {
    //设置正则规则：以数字结尾
    private final static Pattern URLPattern = Pattern
            .compile(".*(\\d+)$");
    //URL满足的条件
    @Override
    public boolean shouldVisit(Page referringPage, WebURL url) {
        String href = url.getURL().toLowerCase();
        return URLPattern.matcher(href).matches()&&(href
            .startsWith("http://the-*****-news.com/news/article"));
    }
    /**
     * 处理URL对应的页面
     */
    @Override
    public void visit(Page page) {
        String url = page.getWebURL().getURL(); // 获取URL
        if (URLPattern.matcher(url).matches() && page.getParseData()
                instanceof HtmlParseData) {
            //解析数据，存储数据
            MYSQLControl.executeInsert(Parse.getData(page));
        }
    }
}
```

JNewsController 为爬虫控制类，如程序 9-11 所示。在该类中，配置了线程数目、是否允许重定向、头信息和超时时间。执行程序 9-11，可以看到数据表 japannews 中成功插入了许多新闻数据，如图 9.8 所示。

程序 9-11

```java
package com.main;

import java.util.HashSet;
import org.apache.http.message.BasicHeader;
import edu.uci.ics.crawler4j.crawler.CrawlConfig;
import edu.uci.ics.crawler4j.crawler.CrawlController;
import edu.uci.ics.crawler4j.fetcher.PageFetcher;
import edu.uci.ics.crawler4j.robotstxt.RobotstxtConfig;
import edu.uci.ics.crawler4j.robotstxt.RobotstxtServer;

public class JNewsController {

    public static void main(String[] args) throws Exception {
        //爬虫状态存储文件夹
        String crawlStorageFolder = "F:/program_work/java_work/CSDNCourse"
                + "/JNewsCrawler/data/craw";
        int numberOfCrawlers = 5;         //线程数
        CrawlConfig config = new CrawlConfig();
        config.setMaxDepthOfCrawling(3);   //只采集第三层页面的数据
        config.setFollowRedirects(false);  //是否允许重定向
        config.setCrawlStorageFolder(crawlStorageFolder);
        HashSet<BasicHeader> collections = new HashSet<BasicHeader>();
        collections.add(new BasicHeader("User-Agent","Mozilla/5.0 "
                + "(Windows NT 10.0;Win64; x64) AppleWebKit/537.36 "
                + "(KHTML, like Gecko) Chrome/63.0.3239.108 "
                + "Safari/537.36"));
        collections.add(new BasicHeader("Accept","text/html,"
                + "application/xhtml+xml,application/xml;q=0.9"
                + ",image/webp,image/apng,*/*;q=0.8"));
        collections.add(new BasicHeader("Accept-Encoding", "gzip, deflate"));
        collections.add(new BasicHeader("Accept-Language", "zh-CN,zh;q=0.9"));
        collections.add(new BasicHeader("Connection", "keep-alive"));
        config.setDefaultHeaders(collections);
        //礼貌采集
        config.setPolitenessDelay(3000);
        config.setSocketTimeout(10000);
```

```
        config.setConnectionTimeout(10000);
        // 配置信息
        PageFetcher pageFetcher = new PageFetcher(config);
        //robots协议
        RobotstxtConfig robotstxtConfig = new RobotstxtConfig();
        RobotstxtServer robotstxtServer = new RobotstxtServer
(robotstxtConfig, pageFetcher);
        CrawlController controller = new CrawlController(config,
                pageFetcher, robotstxtServer);
        //添加种子URL
        controller.addSeed("http://the-*****-news.com/"
                + "news/business");
        controller.addSeed("http://the-*****-news.com/"
                + "news/society");
        controller.addSeed("http://the-*****-news.com/"
                + "news/sports");
        //运行网络爬虫
        controller.start(JNewsCrawler.class, numberOfCrawlers);
    }
}
```

docid	url	time	title	content
4	http://the news.com/news/	2018-11-10T21:09:00+09:00	Caterpillar Japan rep says old-fash	The Japan NewsHarry Kobrak is something of a Japan connoisseur, traveling to local areas of the
5	http://the news.com/news/	2019-07-10T19:08:00+09:00	37% drop in deliveries to cost Boei	ReutersBENGALURU (Reuters) — Boeing Co. is set to lose the title of being the world's biggest pl
6	http://the news.com/news/	2019-06-27T19:29:00+09:00	Rare earths producer secures bette	ReutersMELBOURNE/TOKYO (Reuters) — Australia's Lynas Corp. said on Thursday it signed a 10-
8	http://the news.com/news/	2019-07-02T04:00:00+09:00	Saudi Arabia, Russia set to extend	Reuters VIENNA (Reuters) — OPEC and its allies look set to extend oil supply cuts this week at lea
10	http://the news.com/news/	2019-07-06T19:57:00+09:00	Jaguar Land Rover to build electric	Reuters LONDON (Reuters) — Jaguar Land Rover is making a multi-million pound investment to b
11	http://the news.com/news/	2019-07-07T20:17:00+09:00	Marine insurance premiums surge	The Yomiuri ShimbunConcerns are growing over a possible surge in marine transport costs after
12	http://the news.com/news/	2019-07-07T19:03:00+09:00	Seven-Eleven Japan under fire over	Jiji PressTOKYO (Jiji Press) — Seven-Eleven Japan Co. has been under fire for failing to take suffic
13	http://the news.com/news/	2019-07-10T19:06:00+09:00	Daido Life to launch new product	Jiji Press TOKYO (Jiji Press) — Daido Life Insurance Co. will launch this week a new key person ins
16	http://the news.com/news/	2019-07-06T19:56:00+09:00	Fed faces quandary: to cut rates or	Reuters NEW YORK (Reuters) — U.S. employers are hiring workers, but that is only making the Fe
55	http://the news.com/news/	2019-06-11T18:39:00+09:00	Mobile cancellation fee to be ¥1,00	The Yomiuri ShimbunThe Internal Affairs and Communications Ministry plans to cap the fee for ca
57	http://the news.com/news/	2019-07-08T19:27:00+09:00	Firms develop musical toothbrushe	Jiji PressTOKYO (Jiji Press) — Kyocera Corp., Lion Corp. and Sony Corp. have jointly developed a
58	http://the news.com/news/	2019-07-01T19:51:00+09:00	Govt, firms aim to offer short spac	The Yomiuri ShimbunAiming to realize outer space travel operated by Japanese companies, the g
59	http://the news.com/news/	2019-07-03T18:53:00+09:00	3 can makers to face antitrust fines	Jiji Press TOKYO (Jiji Press) — The Fair Trade Commission plans to fine three major domestic can
107	http://the news.com/news/	2019-06-24T20:32:00+09:00	Nomura's CEO wins vote, survives	BloombergTOKYO (Bloomberg) — Nomura Holdings Inc.'s besieged chief executive won a sharel
109	http://the news.com/news/	2019-06-28T19:44:00+09:00	Cuba to raise state salaries as part	The Associated Press HAVANA (AP) — The Cuban government said Thursday that it is raising state
111	http://the news.com/news/	2019-07-05T19:41:00+09:00	Grannies are delivering ramen on f	BloombergTOKYO (Bloomberg) — Uber Technologies Inc.'s strategy for Japan, where ride-sharin
112	http://the news.com/news/	2019-07-03T18:51:00+09:00	Sony, Daiwa to set up fund for star	Jiji Press TOKYO (Jiji Press) — Sony Corp. and Daiwa Securities Group Inc. said Wednesday they h
154	http://the news.com/news/	2019-07-10T19:10:00+09:00	EU fines Hello Kitty owner $7 millio	The Associated PressBRUSSELS (AP) — European Union authorities on Tuesday fined the compan
155	http://the news.com/news/	2019-06-29T21:24:00+09:00	U.S. delves into Boeing factory issu	Bloomberg CHICAGO (Bloomberg) — The U.S. government has subpoenaed records from a Boei
156	http://the news.com/news/	2019-07-10T19:05:00+09:00	From Nazis to hippies: Beetle's jou	The Associated Press FRANKFURT (AP) — Volkswagen is halting production of the last version of
157	http://the news.com/news/	2019-06-17T20:52:00+09:00	Yamaha puts luxury boats in reach	The Yomiuri ShimbunYamaha Motor Co. is weighing anchor on its boat rental and sharing service
158	http://the news.com/news/	2019-06-27T19:05:00+09:00	Moody's, Israeli firm to make cybe	Reuters TEL AVIV (Reuters) — Ratings agency Moody's Corp. and Israeli cyber group Team8 laun
160	http://the news.com/news/	2019-07-10T19:07:00+09:00	France: No deal with Germany on I	ReutersBERLIN (Reuters) — France is still pushing for a European consensus around a candidate
161	http://the news.com/news/	2019-07-10T19:14:00+09:00	IBM closes $34 bil. deal to buy Rec	Reuters International Business Machines Corp. said Tuesday it has closed its $34 billion acquisitio
162	http://the news.com/news/	2019-07-04T18:47:00+09:00	Otsuka Kagu eyes China market fo	Jiji Press TOKYO (Jiji Press) — Furniture retailer Otsuka Kagu Ltd. will accelerate efforts to develop
165	http://the news.com/news/	2019-07-03T18:54:00+09:00	Govt purse hits record high in fisca	Jiji Press TOKYO (Jiji Press)— The government's general-account tax revenue in fiscal 2018 hit a re

图 9.8　数据表 japannews 中的数据

9.2　WebCollector 的使用

9.2.1　WebCollector 简介

WebCollector 也是一个基于 Java 的开源网络爬虫框架，其支持多线程、深度采集、URL 维护及结构化数据抽取等。WebCollector 项目的源码可以在 GitHub 上进行下载。

相比于 Crawler4j，WebCollector 的可扩展性和适用性更强，如可以实现多代理的切换、断点采集和大规模数据采集。

9.2.2　jar 包的下载

在 Maven 工程的 pom.xml 文件中可以配置 WebCollector 的最新 jar 包以及相应的依赖 jar 包。

```xml
<!-- https://*************.com/artifact/cn.edu.hfut.dmic.
webcollector/WebCollector -->
  <dependency>
    <groupId>cn.edu.hfut.dmic.webcollector</groupId>
    <artifactId>WebCollector</artifactId>
    <version>2.73-alpha</version>
  </dependency>
```

WebCollector 的依赖 jar 包有 OkHttp（网页请求开源库）、Jsoup（HTML/XML 类型的网页解析）、Gson（JSON 数据解析）、slf4j/log4j（配置日志）、rocksdbjni（RocksDB 数据存储）等。

9.2.3　入门案例

使用 WebCollector 仅需要少量的代码便可以开发一款网络爬虫。本案例采集的网站仍是 rediff.com，初始种子 URL 包括如下两个。

```
https://www.******.com              //资讯首页
https://www.******.com/business     //商业资讯
```

与 9.1.3 节的案例相同，需要采集的 URL 的类型需要以"https://www.******.com/"为前缀，以".htm"为后缀。

由 WebCollector 的使用介绍可知，需要创建一个爬虫类 RediffNewsCrawler，该类继承了 BreadthCrawler 类，如程序 9-12 所示。在该程序的构造方法 RediffNewsCrawler()中，使用 addSeed()方法添加种子 URL；使用 addRegex()方法和正则表达式设置 URL 访问规则；并基于传递的参数配置数据保存的文件名和文件编码。在程序采集数据的过程中，种子 URL 也会被访问。为使解析的数据只包含文本类新闻，所以在 visit()方法中添加了待解析 URL 满足的条件。同时，visit()方法调用了文本写入方法——writeFile()

方法,该方法的参数为文件名、需要写入的文本内容和文件编码。在主方法中,使用 setThreads()设置网络爬虫的线程数目;使用 Configuration 类中的 setTopN()方法设置每层最多采集的页面数;使用 start()设置采集的深度,即层数。

程序 9-12

```java
package com.crawler.rediff;

import java.io.BufferedWriter;
import java.io.File;
import java.io.FileOutputStream;
import java.io.IOException;
import java.io.OutputStream;
import java.io.OutputStreamWriter;
import cn.edu.hfut.dmic.webcollector.model.CrawlDatums;
import cn.edu.hfut.dmic.webcollector.model.Page;
import cn.edu.hfut.dmic.webcollector.plugin.rocks.BreadthCrawler;
public class RediffNewsCrawler extends BreadthCrawler {
  private static StringBuilder sb = new StringBuilder();
  private static String fileName;
  private static String code;
  public RediffNewsCrawler(String crawlPath, boolean autoParse,String filename,String cod) {
        super(crawlPath, autoParse);
        /**
         * 添加种子URL
         */
        this.addSeed("https://www.******.com");
        this.addSeed("https://www.******.com/business");
        /**
         * URL访问规则添加
         *
         * 以"https://www.******.com/"为前缀
         * 以".tml"为后缀
         * 不匹配以".(jpg|png|gif|css|js|mid|" + "mp4
         * |wav|avi|mov|mpeg|ram|m4v|pdf)"结尾的URL
         */
        this.addRegex("^(https://www.******.com/).*(\\.htm)$");
        this.addRegex("-.*\\.(jpg|png|gif|css|js|mid|"
```

```java
            + "mp4|wav|avi|mov|mpeg|ram|m4v|pdf)$");
    /**
     * 输出文件配置
     *
     * 文件名及文件编码
     */
    fileName = filename;
    code = cod;
}

public void visit(Page page, CrawlDatums next) {
    String url = page.url();
    //种子URL不符合条件,这里过滤掉
    if (page.matchUrl("^(https://www.******.com"
            + "/).*(\\.htm)$")){
        /**
         *使用Jsoup解析数据
         */
        String title = page.select("#leftcontainer > h1").text();
        String content = page.select("#arti_content_n").text();
        sb.append("URL:\t" + url  + "\n" + "title:\t" + title
                + "\ncontent:\t" + content + "\n\n");
    }
    try {
        writeFile(fileName, sb.toString(), code);
    } catch (IOException e) {
        e.printStackTrace();
    }
}
/**
 * 数据写入指定文档
 *
 * @param file(文件名)
 * @param content(需要写入的内容)
 * @param code(文件编码)
 */
public static void writeFile(String file, String content, String code)
        throws IOException {
```

```
            File result = new File(file);
            OutputStream out = new FileOutputStream(result, false);
            BufferedWriter bw = new BufferedWriter(new OutputStreamWriter
(out, code));
            bw.write(content);
            bw.close();
            out.close();
    }

    public static void main(String[] args) throws Exception {
        RediffNewsCrawler crawler = new RediffNewsCrawler
("rediffNewsCrawler",
                true,"data/rediffNews.txt","utf-8");
        //设置线程数目
        crawler.setThreads(5);
        //设置每一层最多采集的页面数
        crawler.getConf().setTopN(300);
        //开始采集数据,设置采集的深度
        crawler.start(3);
    }
}
```

执行程序 9-12,在控制台会输出一系列日志信息。整个程序的执行时间为 28 秒,一共采集 271 条新闻,采集到数据写入了项目 "data/" 目录下的文件 "rediffNews.txt",其内容如图 9-9 所示。相比 Crawler4j,WebCollector 采集数据的速度更快、效率更高。在执行的过程中,在项目的根目录下会自动创建文件夹 "rediffNewsCrawler",同时在该文件夹下还包含三个文件,分别是 "crawldb"、"fetch" 和 "link"。这三个文件夹下存放的都是 RocksDB 数据库对应的文件,如图 9.10 所示。RocksDB 是一种嵌入式的支持持久化的 key-value 存储系统,使用程序 9-13,可以在控制台输出每个数据库中存储的 key-value 数据。其中,key 为访问的 URL;value 为包含 5 个字段的 JSON 数据,即 URL、状态 status(三种取值为 0、1、5,默认为 0)、执行时间 executeTime(UNIX 时间戳-毫秒)、执行次数 executeCount(默认值为 0)、HTTP 状态码(默认值为-1,表示未获取到状态码)和重定向地址(默认值为 null,如果有重定向则保存重定向地址)。WebCollector 的数据去重,依据的是 RocksDB 数据库中的 key,如果没有设置 key,程序会将 URL 当成 key。

图 9.9 文件"rediffNews.txt"中的数据

图 9.10 RocksDB 数据库对应的文件

程序 9-13

```java
package com.crawler.rediff;

import org.rocksdb.Options;
import org.rocksdb.RocksDB;
import org.rocksdb.RocksIterator;
public class RocksDBOpen {
    static{
        RocksDB.loadLibrary();
    }
    static RocksDB rocksDB;
    static String path = "rediffNewsCrawler/crawldb";
    // static String path = "rediffNewsCrawler/fetch";
    // static String path = "rediffNewsCrawler/link";
    public static void main(String[] args) throws Exception {
        Options options = new Options();
        options.setCreateIfMissing(true);
        //打开RocksDB
```

```java
rocksDB = RocksDB.open(options, path);
//迭代, 输出内容
RocksIterator iter = rocksDB.newIterator();
for(iter.seekToFirst(); iter.isValid(); iter.next()) {
    System.out.println("key:" + new String(iter.key()) +
            ",value:" + new String(iter.value()));
}
    }
}
```

9.2.4 相关配置

在程序 9-12 中，配置了网络爬虫的线程数目、每层最多采集的页面数和采集深度。下面，将介绍 WebCollector 的其他相关配置。

1. User-Agent

在 cn.edu.hfut.dmic.webcollector.util 包的 Config 类中，给出了 User-Agent 的默认值，即：

```
String DEFAULT_USER_AGENT = "Mozilla/5.0 (X11; Ubuntu; Linux x86_64; rv:36.0) Gecko/20100101 Firefox/36.0";
```

而在 cn.edu.hfut.dmic.webcollector.conf 包的 Configuration 类中，提供了配置 User-Agent 的方法，即 setDefaultUserAgent()方法，其使用方式如下所示。

```
crawler.getConf().setDefaultUserAgent("Your USER_AGENT");
```

2. 请求时间间隔

为了礼貌地采集网站数据，Configuration 类提供了 setExecuteInterval ()方法来设置任意线程 URL 请求之间的时间间隔（默认值为 0）。以下为配置程序。

```
crawler.getConf().setExecuteInterval(1000);
```

3. 超时时间

WebCollector 提供了连接超时时间和获取数据超时时间的配置。在 Config 类中，连接超时时间的默认值为 3 秒，获取数据超时时间的默认值为 10 秒。配置这两种超时时间，可以使用 Configuration 类中的 setConnectTimeout()和 setReadTimeout()方法。

```
crawler.getConf().setConnectTimeout(10000);  //连接超时
crawler.getConf().setReadTimeout(20000);     //获取数据超时
```

4. 最大重定向次数

在 WebCollector 中，最大重定向次数默认设置为 2 次。但使用者可以使用 Configuration 类中的 setMaxRedirect()重新配置。

```
crawler.getConf().setMaxRedirect(5);
```

5. 最大执行次数

使用 Crawler 类中 setMaxExecuteCount()可以设置爬虫任务的最大执行次数。在数据采集任务中，请求 URL 和解析数据出错都有可能导致任务失败。当某个任务执行失败时，如果设置的最大执行次数超过 1，那么该任务还会重新执行，直到达到最大执行次数。setMaxExecuteCount()方法的默认设置为-1，即任务失败不会重新执行，以下为该方法的使用方式。

```
crawler.setMaxExecuteCount(2);
```

6. 断点爬取

WebCollector 的一个重要特性便是支持断点采集。针对耗时较长和大规模数据采集的任务，经常会遇到一些意外情况（如断网、死机和断电等），导致程序中断。为了保证数据采集任务不受这些因素的影响，WebCollector 框架中提供了断点配置方法——setResumable()方法，该方法的参数类型为 boolean 类型。默认情况下，setResumable()方法输入参数设置为 false，即每次采集清空历史数据，不执行断点爬取。如程序 9-12 中没有设置断点爬取，则每次启动任务"crawldb"、"fetch"和"link"三个数据库中的数据都会被清空。但如果在程序 9-12 中的 start()方法之前进行如下配置：

```
crawler.setResumable(true);
```

再次执行程序 9-12 后，同时使用程序 9-13 打开"crawldb"数据库，则会发现其原有的 key-value 数据依旧还在，并且添加了新的 key-value 数据。

9.2.5　HTTP 请求扩展

WebCollector 2.7 版本默认使用 cn.edu.hfut.dmic.webcollector.plugin.net 包中的 OkHttpRequester 作为 HTTP 请求插件，但其提供的功能有限，为进一步扩展 HTTP 请求的功能（如设置 HTTP 请求头信息、设置代理、设置请求方法等），可继承 OkHttpRequester 类，复写其中的 createRequestBuilder()方法。程序 9-14 采集的数据仍是 rediff.com 中的新闻。相比于程序 9-12，这里自定义了请求插件，即 MyRequester，其继承了 OkHttpRequester。并且，在复写的 createRequestBuilder()方法中使用了 addHeader()方法添加请求头信息。在构造方法 HeaderAdd ()中，只要使用 setRequester() 方法配置自定义的 MyRequester，便可以利用自定义的请求插件请求 URL。

程序 9-14

```java
package com.crawler.rediff;

import java.io.BufferedWriter;
import java.io.File;
import java.io.FileOutputStream;
import java.io.IOException;
import java.io.OutputStream;
import java.io.OutputStreamWriter;
import cn.edu.hfut.dmic.webcollector.model.CrawlDatum;
import cn.edu.hfut.dmic.webcollector.model.CrawlDatums;
import cn.edu.hfut.dmic.webcollector.model.Page;
import cn.edu.hfut.dmic.webcollector.plugin.net.OkHttpRequester;
import cn.edu.hfut.dmic.webcollector.plugin.rocks.BreadthCrawler;
import okhttp3.Request;

/**
 * 根据WebCollector作者的案例改编
 */
public class HeaderAdd extends BreadthCrawler {
    private static StringBuilder sb = new StringBuilder();
    private static String fileName;
    private static String code;
    // 自定义请求头
    public static class MyRequester extends OkHttpRequester {
        //每次发送请求前都会使用这个方法来构建请求
        @Override
        public Request.Builder createRequestBuilder(CrawlDatum crawlDatum) {
            //使用的是OkHttp中的Request.Builder
            return super.createRequestBuilder(crawlDatum)
                    .addHeader("User-Agent","Mozilla/5.0 (Windows"
                            + " NT 10.0; Win64; x64) AppleWebKit/5"
                            + "37.36 (KHTML, like Gecko) Chrome/63"
                            + ".0.3239.108 Safari/537.36")
                    .addHeader("Accept", "text/html,application/"
```

```java
                    + "xhtml+xml,application/xml;q=0.9,"
                    + "image/webp,image/apng,*/*;q=0.8")
                .addHeader("Connection", "keep-alive");
        }
    }

    public HeaderAdd(String crawlPath, boolean autoParse,String filename,String cod) {
        super(crawlPath, autoParse);
        setRequester(new MyRequester());
        /**
         * 添加种子URL
         */
        this.addSeed("https://www.******.com");
        this.addSeed("https://www.******.com/business");
        /**
         * URL访问规则添加
         *
         * 以 "https://www.******.com/" 为前缀
         * 以 ".tml" 为后缀
         * 不匹配以 ".(jpg|png|gif|css|js|mid)" + "mp4
         * |wav|avi|mov|mpeg|ram|m4v|pdf)" 结尾的URL
         */
        this.addRegex("^(https://www.******.com/).*(\\.htm)$");
        this.addRegex("-.*\\.(jpg|png|gif|css|js|mid|"
                + "mp4|wav|avi|mov|mpeg|ram|m4v|pdf)$");
        /**
         * 输出文件配置
         *
         * 文件名及文件编码
         */
        fileName = filename;
        code = cod;
    }
    public void visit(Page page, CrawlDatums crawlDatums) {
        String url = page.url();
        //种子URL不符合条件，这里过滤掉
        if (page.matchUrl("^(https://www.******.com"
                + "/).*(\\.htm)$")){
            /**
             *使用Jsoup解析数据
             */
```

```java
            String title = page.select("#leftcontainer > h1").text();
            String content = page.select("#arti_content_n").text();
            sb.append("URL:\t" + url + "\n" + "title:\t" + title
                    + "\ncontent:\t" + content + "\n\n");
        }
        try {
            writeFile(fileName, sb.toString(), code);
        } catch (IOException e) {
            e.printStackTrace();
        }
    }
    /**
     * 数据写入指定文档
     *
     * @param file(文件名)
     * @param content(需要写入的内容)
     * @param code(文件编码)
     */
    public static void writeFile(String file, String content, String code)
            throws IOException {
        File result = new File(file);
        OutputStream out = new FileOutputStream(result, false);
        BufferedWriter bw = new BufferedWriter(new OutputStreamWriter(out, code));
        bw.write(content);
        bw.close();
        out.close();
    }
    public static void main(String[] args) throws Exception {
        HeaderAdd crawler = new HeaderAdd("rediffNewsCrawler_head", true,
                "data/rediffNews.txt","utf-8");
        //设置线程数目
        crawler.setThreads(10);
        //设置每一层最多采集的页面数
        crawler.getConf().setTopN(400);
        //礼貌采集
        crawler.getConf().setExecuteInterval(1000);
        //开始采集数据,设置采集的深度
        crawler.start(4);
```

　　　　}
　　}

另外，通过复写 createRequestBuilder()方法，也可以实现表单数据的提交。以 4.1.4 节的快递单号查询为例，程序 9-15 为 WebCollector 实现表单数据提交的代码。WebCollector 的一个优点是可以为每个采集任务配置附加信息，如使用 CrawlDatum 类中的 meta(String key, int value)方法添加一系列的 key-value 类型的参数。在复写的方法中，可以使用 meta(String key)方法轻松捕获该采集任务的附加信息。

程序 9-15

```java
package com.crawler.sina;
import okhttp3.Request;
import cn.edu.hfut.dmic.webcollector.model.CrawlDatum;
import cn.edu.hfut.dmic.webcollector.model.CrawlDatums;
import cn.edu.hfut.dmic.webcollector.model.Page;
import cn.edu.hfut.dmic.webcollector.plugin.net.OkHttpRequester;
import cn.edu.hfut.dmic.webcollector.plugin.rocks.BreadthCrawler;
import okhttp3.MultipartBody;
import okhttp3.RequestBody;
/**
 * 根据WebCollector作者的案例改编
 *
 * 提交表单数据
 */
public class PostRequestTest extends BreadthCrawler {
  public PostRequestTest(final String crawlPath, boolean autoParse) {
      super(crawlPath, autoParse);
      addSeed(new CrawlDatum("http://www.*****.com/ems.php")
              .meta("wen", "EH629625211CS")
              .meta("action", "ajax"));
      setRequester(new OkHttpRequester(){
          @Override
          public Request.Builder createRequestBuilder(CrawlDatum crawlDatum) {
              Request.Builder requestBuilder = super.createRequestBuilder(crawlDatum);
              RequestBody requestBody;
              String wen = crawlDatum.meta("wen");
              // 如果没有表单数据
              if(wen == null){
                  requestBody = RequestBody.create(null, new
```

```
            byte[]{});
                    }else{
                        //根据meta创建请求体
                        requestBody = new MultipartBody.Builder()
                                .setType(MultipartBody.FORM)
                                .addFormDataPart("wen", wen)
                                .addFormDataPart("action", crawlDatum.
meta("action"))
                                .build();
                    }
                    return requestBuilder.post(requestBody)
                            .header("Connection", "keep-alive");
                }
            });
        }
        public void visit(Page page, CrawlDatums next) {
            String html = page.html();
            System.out.println("快递信息" + html);
        }

        public static void main(String[] args) throws Exception {
            PostRequestTest crawler = new PostRequestTest
("post_crawler", true);
            crawler.start(1);
        }
    }
```

在程序 9-14 和 9-15 中，构建 HTTP 请求插件使用的是 OkHttp jar 包中的 Request 类的内部类 Builder。但这个 Builder 类只能用于配置请求方法（GET/POST）、请求头（添加/删除）和请求体等。如果要配置代理和超时时间等内容，则需要使用 OkHttpClient 类中的内部类 Builder。在程序 9-16 中，实现了 HTTP 请求的多代理随机切换模式。MyRequester 类继承了 OkHttpRequester 类，但复写的是 createOkHttpClientBuilder() 方法。在该方法中，使用了 OkHttpClient.Builder 类中的 proxySelector() 添加代理。在采集的过程中，一些代理可能失效，导致 URL 请求失败，为此，在主方法中，设置了断点采集。

程序 9-16

```
package com.crawler.rediff;

import cn.edu.hfut.dmic.webcollector.model.CrawlDatums;
```

```java
import cn.edu.hfut.dmic.webcollector.model.Page;
import cn.edu.hfut.dmic.webcollector.net.Proxies;
import cn.edu.hfut.dmic.webcollector.plugin.net.OkHttpRequester;
import cn.edu.hfut.dmic.webcollector.plugin.rocks.BreadthCrawler;
import okhttp3.OkHttpClient;
import java.io.IOException;
import java.net.Proxy;
import java.net.ProxySelector;
import java.net.SocketAddress;
import java.net.URI;
import java.util.ArrayList;
import java.util.List;

/**
 * 根据WebCollector作者的案例改编
 *
 */

public class ProxyUseTest extends BreadthCrawler {
    /**
     * 自定义的请求插件
     * 添加多个代理
     * 使用代理选择器，实现随机代理切换
     */
    public static class MyRequester extends OkHttpRequester {
        Proxies proxies;
        public MyRequester() {
            proxies = new Proxies();
            proxies.addSocksProxy("127.0.0.1", 1080); //本机
            proxies.addSocksProxy("183.161.29.127", 8060);
            proxies.addSocksProxy("163.125.248.171", 8118);
            // 直接连接，不使用代理
            proxies.add(null);
        }
        @Override
        public OkHttpClient.Builder createOkHttpClientBuilder() {
            return super.createOkHttpClientBuilder()
                    // 设置一个代理选择器
                    .proxySelector(new ProxySelector() {
                        @Override
                        public List<Proxy> select(URI uri) {
                            //随机选择一个代理
```

```java
                        Proxy randomProxy = proxies.randomProxy();
                        //返回值类型需要为List
                        List<Proxy> randomProxies = new ArrayList<Proxy>();
                        //如果随机到null,即不需要代理,返回空的List即可
                        if(randomProxy != null) {
                            randomProxies.add(randomProxy);
                        }
                        System.out.println("使用的代理为:" + randomProxies);
                        return randomProxies;
                    }
                    @Override
                    public void connectFailed(URI uri,
                            SocketAddress sa, IOException ioe) {

                    }
                });
        }
    }
    public ProxyUseTest(String crawlPath) {
        super(crawlPath, true);
        // 设置请求插件
        setRequester(new MyRequester());
        // 采集新闻
        this.addSeed("https://www.******.com");
        this.addSeed("https://www.******.com/business");
        this.addRegex("^(https://www.******.com/).*(\\.htm)$");

    }

    public void visit(Page page, CrawlDatums crawlDatums) {
        if (page.matchUrl("^(https://www.******.com/"
                + ").*(\\.htm)$")){
            String title = page
                    .select("#leftcontainer > h1").text();
            System.out.println("标题为:" + title);
        }

    }
```

```
    public static void main(String[] args) throws Exception {
        ProxyUseTest crawler = new ProxyUseTest
("crawl_proxy_rediff");
        //设置线程数目
        crawler.setThreads(3);
        //防止有些代理不可用，下次启动可以使用其他代理继续请求
        crawler.setResumable(false);
        //礼貌采集
        crawler.getConf().setExecuteInterval(1000);
        //设置每一层最多采集的页面数
        crawler.getConf().setTopN(100);
        crawler.start(3);
    }
}
```

9.2.6 翻页数据采集

在程序 9-15 中，已介绍了 meta() 方法的特性，即为每次的数据采集任务添加附加信息。灵活运用 meta() 方法的特性可以简化爬虫设计，如添加请求参数、标记页码和标记采集的层次等。另外，在 CrawlDatum 类中，type(String type) 方法调用了 meta(String key, String value) 方法，并将 key 设置为 "s_t"，该方法常用来标记每个采集任务的类型。例如，在采集具有多页的新闻页面时，可以调用 type() 方法将每页的新闻数据设置为 "firstLayer"，即新闻标题页；同时，将该页面中链接的每一条新闻 URL 的 type 设置为 content，即新闻详情页。下面，将利用 CrawlDatum 类的这两种方法，实现翻页数据的采集。采集的网站为某大学新闻网，如图 9.11 所示。

图 9.11　某大学新闻网主页

程序 9-17 给出了完整的数据采集代码。在构造方法中，使用了 addSeed(CrawlDatum datum) 方法来添加种子节点，每个种子节点相当于一个页面，这里使用循环的方式实现翻页操作。在实例化 CrawlDatum 时，使用了 type() 方法以及 meta(String key, String value)方法为每个待采集的种子节点配置附加信息。在程序 9-17 的 visit()方法中，首先判断页面是新闻标题页还是新闻详情页。如果是新闻标题页，则解析得到新闻 URL、新闻标题、新闻时间，并将数据写入指定文本；同时，利用 CrawlDatum 类中的 addAndReturn()方法将新闻解析的 URL 添加到后续采集任务中，并利用 type()方法将这些 URL 的类型设置为"content"，利用 meta(String key, String value)方法标记新闻所在的页面和序号。如果是新闻详情页，则输出新闻的 URL、新闻所在的页面编号、新闻所在页面中的序号和新闻的详细内容。执行该程序，会在工程的"data"目录下，产生"hfut_newsUrl.txt"和"hfut_newsContent.txt"两个文件，这两个文件的部分数据分别如图 9.12 和图 9.13 所示。

程序9-17

```java
package com.crawler.hfutnews;
import cn.edu.hfut.dmic.webcollector.model.CrawlDatum;
import cn.edu.hfut.dmic.webcollector.model.CrawlDatums;
import cn.edu.hfut.dmic.webcollector.model.Page;
import cn.edu.hfut.dmic.webcollector.plugin.ram.RamCrawler;
import org.jsoup.select.Elements;
import java.io.BufferedWriter;
import java.io.File;
import java.io.FileOutputStream;
import java.io.IOException;
import java.io.OutputStream;
import java.io.OutputStreamWriter;
import org.jsoup.nodes.Element;
/**
 *
 * 这里使用RamCrawler
 * RamCrawler不需要依赖文件系统或数据库，适合一次性的爬取任务
 * 也可将该程序改写为BreadthCrawler爬虫
 */
public class HFUTNewsCrawler extends RamCrawler {
    String fileFirstLayerOutPut = "data/hfut_newsUrl.txt";
    String contentOutPut = "data/hfut_newsContent.txt";
    String code = "utf-8";
    StringBuilder sb_first = new StringBuilder();
    StringBuilder sb_content = new StringBuilder();
    public HFUTNewsCrawler(int pageNum) throws Exception {
        //添加多页
```

```java
        for (int pageIndex = 1; pageIndex <= pageNum; pageIndex++) {
            String url = "http://news.****.edu.cn/list-1-" + pageIndex + ".html";
            CrawlDatum datum = new CrawlDatum(url)
                    .type("firstLayer")              //第一层
                    .meta("pageIndex", pageIndex)    //页面保存
                    .meta("depth", 1);               //深度为第一层
            this.addSeed(datum);
        }
    }
    public void visit(Page page, CrawlDatums next) {
        int pageIndex = page.metaAsInt("pageIndex");
        int depth = page.metaAsInt("depth");
        if (page.matchType("firstLayer")) {
            //解析新闻标题页
            Elements results = page.select("div.col-lg-8 > ul").select("li");
            for (int rank = 0; rank < results.size(); rank++) {
                Element result = results.get(rank);
                String href = "http://news.****.edu.cn" +
                        result.select("a").attr("href");
                String title = result.select("a").text();
                String time = result.select("span[class=rt]").text();
                if (title.length() != 0) {
                    //输出第一层信息
                    sb_first.append("url:" + href + "\ttitle:" + title +
                            "\ttime:" + time + "\n");
                    try {
                        writeFile(fileFirstLayerOutPut, sb_first.toString(), code);
                    } catch (IOException e) {
                        e.printStackTrace();
                    }
                    /*
                     * 添加需要访问的新闻链接,类型为content
                     *
                     * 用于爬取新闻的详细内容
                     */
                    //将该URL添加到CrawlDatum作为要采集的URL
```

```java
                        next.addAndReturn(href)
                            .type("content")              //内容页面
                            .meta("pageIndex", pageIndex) //第几页的新闻
                            .meta("rank", rank);          //这条新闻的序号
                    }
                }
            }
            //新闻详情页
            if (page.matchType("content")) {
                //输出结果
                String url = page.url();
                int Index = page.metaAsInt("pageIndex"); //新闻在第几页
                int rank = page.metaAsInt("rank"); //新闻在页面的序号
                //新闻内容
                String content = page.select("div[id=artibody]").text();
                //输出第二层信息
                sb_content.append("url:" + url + "\tIndex:" + Index +
"\trank:" + rank +
                        "\tcontent:" + content + "\n");
                try {
                    writeFile(contentOutPut, sb_content.toString(), code);
                } catch (IOException e) {
                    e.printStackTrace();
                }
            }
            /*
             * 页面的深度+1
             *
             * 新闻的详情页
             */
            next.meta("depth", depth + 1);
        }
        /**
         * 数据写入指定文档
         *
         * @param file(文件名)
         * @param content(需要写入的内容)
         * @param code(文件编码)
         */
```

```
    public static void writeFile(String file, String content, String code)
            throws IOException {
        File result = new File(file);
        OutputStream out = new FileOutputStream(result, false);
        BufferedWriter bw = new BufferedWriter(new OutputStreamWriter
(out, code));
        bw.write(content);
        bw.close();
        out.close();
    }
    public static void main(String[] args) throws Exception {
        //添加爬取的页面数
        HFUTNewsCrawler crawler = new HFUTNewsCrawler(3);
        //添加线程数
        crawler.setThreads(10);
        //启动采集程序
        crawler.start();
    }
  }
```

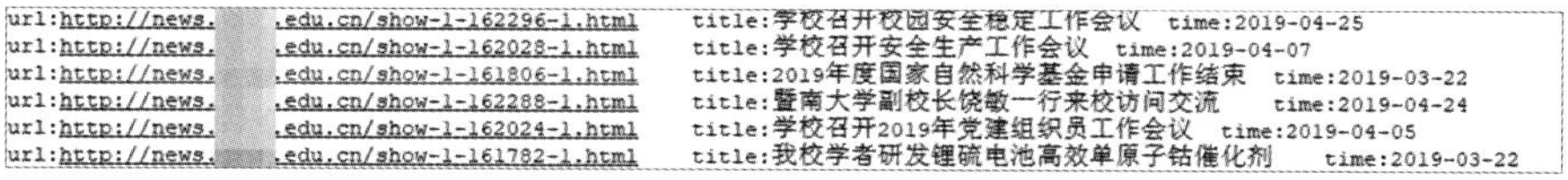

```
url:http://news.[illegible].edu.cn/show-1-162296-1.html    title:学校召开校园安全稳定工作会议  time:2019-04-25
url:http://news.[illegible].edu.cn/show-1-162028-1.html    title:学校召开安全生产工作会议  time:2019-04-07
url:http://news.[illegible].edu.cn/show-1-161806-1.html    title:2019年度国家自然科学基金申请工作结束  time:2019-03-22
url:http://news.[illegible].edu.cn/show-1-162288-1.html    title:暨南大学副校长饶敏一行来校访问交流  time:2019-04-24
url:http://news.[illegible].edu.cn/show-1-162024-1.html    title:学校召开2019年党建组织员工作会议  time:2019-04-05
url:http://news.[illegible].edu.cn/show-1-161782-1.html    title:我校学者研发锂硫电池高效单原子钴催化剂  time:2019-03-22
```

图 9.12 “hfut_newsUrl.txt”文件中的部分数据

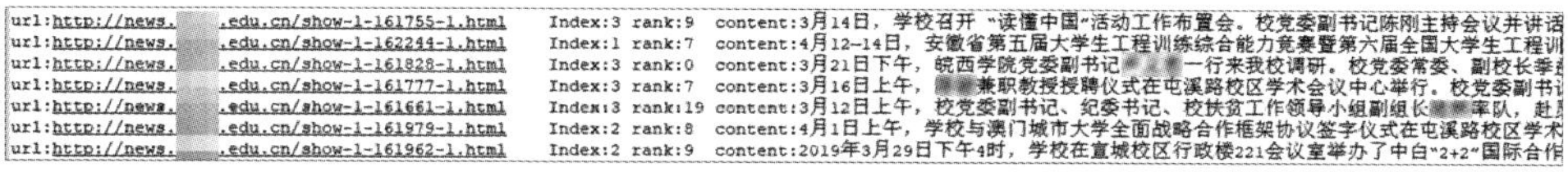

```
url:http://news.[illegible].edu.cn/show-1-161755-1.html    Index:3 rank:9  content:3月14日，学校召开“读懂中国”活动工作布置会。校党委副书记陈刚主持会议并讲话
url:http://news.[illegible].edu.cn/show-1-162244-1.html    Index:1 rank:7  content:4月12-14日，安徽省第五届大学生工程训练综合能力竞赛暨第六届全国大学生工程训
url:http://news.[illegible].edu.cn/show-1-161828-1.html    Index:3 rank:0  content:3月21日下午，皖西学院党委副书记[illegible]一行来我校调研。校党委常委、副校长季
url:http://news.[illegible].edu.cn/show-1-161777-1.html    Index:3 rank:7  content:3月16日上午，[illegible]兼职教授授聘仪式在屯溪路校区学术会议中心举行。校党委副书记
url:http://news.[illegible].edu.cn/show-1-161661-1.html    Index:3 rank:19 content:3月12日上午，校党委副书记、纪委书记、校扶贫工作领导小组副组长[illegible]率队，赴
url:http://news.[illegible].edu.cn/show-1-161979-1.html    Index:2 rank:8  content:4月1日上午，学校与澳门城市大学全面战略合作框架协议签字仪式在屯溪路校区学术
url:http://news.[illegible].edu.cn/show-1-161962-1.html    Index:2 rank:9  content:2019年3月29日下午4时，学校在宣城校区行政楼221会议室举办了中白“2+2”国际合作
```

图 9.13 “hfut_newsContent.txt”文件中的部分数据

9.2.7 图片的采集

本节将以 Picjumbo 图片网为例，介绍如何使用 WebCollector 采集图片数据。用户进入网站首页，可以在搜索框中搜索指定关键词(如“macbook”)，单击搜索，网页会返回关键词对应的图片，如图 9.14 所示。由图 9.14 可知，一个关键词会对应多页的内容。本节的目标是将指定关键词对应的多页图片数据，保存到指定文件夹。

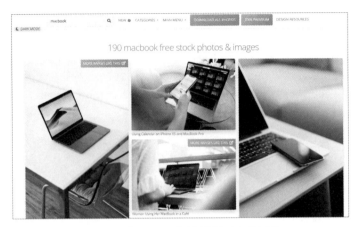

图 9.14 "macbook"关键词对应的图片

程序 9-18 给出了图片采集的完整代码。在构造方法中，首先拼接关键词对应的每页 URL，并添加为种子节点，之后使用正则表达式限定数据采集范围。在 visit()方法中，使用响应头 Content-Type 信息来判断当前资源对应的网页是否为图片，如果为图片则保存到指定目录下。在主方法中，设置要爬取的关键词和页数。同时，利用 setResumable()方法配置断点采集，利用 setAutoDetectImg()方法设置自动解析图片链接，利用 setMaxReceiveSize()方法设置网页内容的容量上限。执行程序 9-18，会发现在工程根目录的"images"文件夹下，保存了一系列的图片，如图 9.15 所示。

程序 9-18

```
package com.crawler.picture;

import cn.edu.hfut.dmic.webcollector.model.CrawlDatums;
import cn.edu.hfut.dmic.webcollector.model.Page;
import cn.edu.hfut.dmic.webcollector.plugin.berkeley.BreadthCrawler;
import cn.edu.hfut.dmic.webcollector.util.ExceptionUtils;
import cn.edu.hfut.dmic.webcollector.util.FileUtils;
import cn.edu.hfut.dmic.webcollector.util.MD5Utils;
import java.io.File;
import java.io.UnsupportedEncodingException;
/**
 * 采集关键词相关的图片
 *
 *
 */
public class ImageCrawler extends BreadthCrawler {
    //文件保存的目录
```

```java
        File baseDir = new File("images");
        public ImageCrawler(String crawlPath,String keyWord, int pageNum)
throws UnsupportedEncodingException {
            super(crawlPath, true);
            //添加种子URL
            for (int i = 1; i <= pageNum; i++) {
                String url = "https://*******.com/page/" + i +
"/?s=macbook";
                this.addSeed(url);
            }
            //限定爬取范围
            this.addRegex
("^(https://*******.com/wp-content/uploads).*(\\.jpg)$");
        }
        public void visit(Page page, CrawlDatums next) {
            String contentType = page.contentType();
            //根据Http头中的Content-Type信息来判断当前资源是网页还是图片
            if(contentType!=null && contentType.startsWith("image")){
                //从Content-Type中获取图片扩展名
                String extensionName = contentType.split("/")[1];
                try {
                    byte[] image = page.content();
                    //根据图片MD5生成文件名
                    String fileName = String.format("%s.%s",
                            MD5Utils.md5(image), extensionName);
                    File imageFile = new File(baseDir, fileName);
                    FileUtils.write(imageFile, image);
                } catch (Exception e) {
                    ExceptionUtils.fail(e);
                }
            }
        }

        public static void main(String[] args) throws Exception {
            ImageCrawler imageCrawler = new ImageCrawler
("crawl_image","macbook",4);
            //设置为断点爬取,否则每次开启爬虫都会重新爬取
            //imageCrawler.setResumable(true);
            //设置自动解析图片链接
            imageCrawler.getConf().setAutoDetectImg(true);
            /*使用默认的Requester,需要设置网页大小上限
```

```
     *  否则可能会获得一个不完整的页面
     *  如接收页面大小上限设置为10M
     */
    imageCrawler.getConf().setMaxReceiveSize(1024 * 1024 * 10);
    imageCrawler.setThreads(30);
    imageCrawler.start(4);
  }
}
```

图 9.15　目录 "images" 文件夹下保存的图片

9.2.8　数据采集入库

本节将使用 WebCollector 采集 makro.co.za（国外某电商平台）中电子产品相关的数据。数据采集共分为三层次，第一层次为品牌页，如图 9.16 所示；第二层次为品牌对应的产品页，如图 9.17 所示；第三层次为产品详情页，如图 9.18 所示。

图 9.16　品牌页

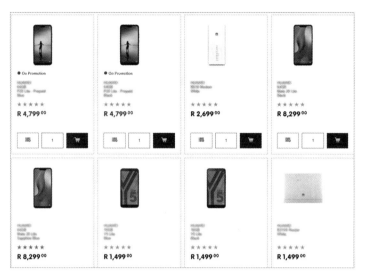

图 9.17 品牌对应的产品页

图 9.18 产品详情页

品牌页作为种子节点，需要将其 type（类型）设置为 FirstLayer。解析品牌页，获取第二层次的相关 URL，即品牌对应的产品页入口，并将其 type 设置为 SecondLayer。解析品牌对应的产品页，获取第三层次的相关 URL，即产品详情页，并将其 type 设置为 ThirdLayer。

本案例需要存储的数据包括品牌对应的产品信息（第二层次的信息）和产品详情信息（第三层次的信息）。

1. 框架内容

本项目依旧使用第 7 章介绍的框架结构编写，框架中的 package 有 com.(db、main、model、parse)。图 9.19 为整个项目的结构。

编写项目之前，使用 Maven 工程中的 pom.xml 文件配置整个项目需要的 jar 包。

```xml
<!-- https://************.com/artifact/cn.edu.hfut.
dmic.webcollector/WebCollector -->
    <dependency>
        <groupId>cn.edu.hfut.dmic.webcollector</groupId>
        <artifactId>WebCollector</artifactId>
        <version>2.73-alpha</version>
    </dependency>
    <!-- 数据库相关 -->
    <dependency>
        <groupId>mysql</groupId>
        <artifactId>mysql-connector-java</artifactId>
        <version>5.1.32</version>
    </dependency>
    <dependency>
        <groupId>commons-dbutils</groupId>
        <artifactId>commons-dbutils</artifactId>
        <version>1.7</version>
    </dependency>
    <dependency>
        <groupId>org.apache.commons</groupId>
        <artifactId>commons-dbcp2</artifactId>
        <version>2.5.0</version>
    </dependency>
    <!-- Json 解析 Fastjson -->
    <dependency>
        <groupId>com.alibaba</groupId>
        <artifactId>fastjson</artifactId>
        <version>1.2.47</version>
    </dependency>
    <!-- Json解析Fastjson -->
    <dependency>
        <groupId>com.alibaba</groupId>
        <artifactId>fastjson</artifactId>
        <version>1.2.47</version>
    </dependency>
```

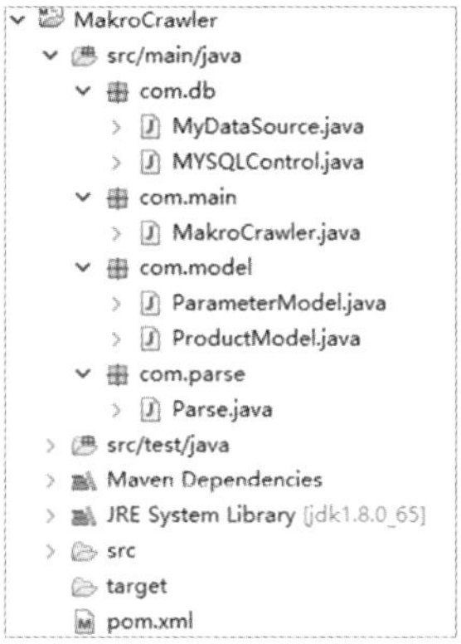

```
          MakroCrawler
          src/main/java
              com.db
                  MyDataSource.java
                  MYSQLControl.java
              com.main
                  MakroCrawler.java
              com.model
                  ParameterModel.java
                  ProductModel.java
              com.parse
                  Parse.java
          src/test/java
          Maven Dependencies
          JRE System Library [jdk1.8.0_65]
          src
          target
          pom.xml
```

图 9.19 项目结构

2．程序设计

（1）com.model

品牌对应的产品信息包括：产品 id、产品名称、产品价格、产品对应的品牌和产品的类别。为此，在 com.model 下构建 ProductModel 类用于封装要存储的数据，如程序 9-19 所示。产品详情信息包括产品 id、产品描述和产品型号，为此，在 com.model 下又构建 ParameterModel，如程序 9-20 所示。

程序 9-19

```java
package com.model;

public class ProductModel {
    private String id;          //产品id
    private String name;        //产品名称
    private String price;       //产品价格
    private String brand;       //产品对应的品牌
    private String category;    //产品的类别
    public String getId() {
        return id;
    }
    public void setId(String id) {
        this.id = id;
    }
    public String getName() {
        return name;
    }
    public void setName(String name) {
        this.name = name;
```

```java
    }
    public String getPrice() {
        return price;
    }
    public void setPrice(String price) {
        this.price = price;
    }
    public String getBrand() {
        return brand;
    }
    public void setBrand(String brand) {
        this.brand = brand;
    }
    public String getCategory() {
        return category;
    }
    public void setCategory(String category) {
        this.category = category;
    }
}
```

程序 9-20

```java
package com.model;

public class ParameterModel {
    private String product_id;       //产品id
    private String description;      //产品描述
    private String model;            //产品型号
    public String getProduct_id() {
        return product_id;
    }
    public void setProduct_id(String product_id) {
        this.product_id = product_id;
    }
    public String getDescription() {
        return description;
    }
    public void setDescription(String description) {
        this.description = description;
    }
    public String getModel() {
        return model;
```

```java
    }
    public void setModel(String model) {
        this.model = model;
    }
}
```

(2) com.parse

程序 9-21 为解析类 Parse。其中，getProductData() 方法负责解析品牌对应的产品信息。在解析数据之前，使用浏览器抓包分析待解析的字段存放的形式和位置，发现产品相关的数据都存放到 HTML 中的 input 标签中，并且以 JSON 形式呈现，如

```
<input type='hidden' class='js-gtmProductListItem' value='{"name":
"Capdase Alumor Blackberry 9360 Cover (Dark Pink)","id":
"091a30a2-88cf-4890-9c94-2a92be71623c","price":"99.0","brand":
"Capdase","category":"Cell Pouches","variant":""}'/>
```

由于每个品牌对应的页面（见图 9.17）包含多个产品，因此在解析的过程中需要将每个产品的信息封装到实例化的 ProductModel 中，并将其添加到集合 listPro 中。

另外，为获取产品的详细参数，还需将解析得到的产品详情页 URL 添加为待访问的页面，并将其 type 设置为 ThirdLayer。getParData() 方法负责解析产品的详细参数，即产品描述和产品型号。

程序 9-21

```java
package com.parse;

import java.util.List;
import org.jsoup.nodes.Element;
import org.jsoup.select.Elements;
import com.alibaba.fastjson.JSON;
import com.model.ParameterModel;
import com.model.ProductModel;
import cn.edu.hfut.dmic.webcollector.model.CrawlDatums;
import cn.edu.hfut.dmic.webcollector.model.Page;

public class Parse {
    /**
     * 针对每个品牌
     * 解析得到品牌下产品的id、名称等
     * 将产品的URL添加到下一层需要访问的页面
     * 并将type设置为ThirdLayer
```

```java
     */
    public static void getProductData(Page page, List<ProductModel>
listPro, CrawlDatums next) {
        //解析第二层页面
        Elements elements = page
                .select("input[class=js-gtmProductListItem]");
        for (Element ele : elements) {
            //数据封装在JSON中,使用Fastjson解析
            ProductModel model = JSON.parseObject(ele.attr("value"),
                    ProductModel.class);
            listPro.add(model);
        }
        //添加第三层需要采集的URL
        Elements urlElements = page
                .select("a[class=product-tile-inner__img]");
        for (Element ele : urlElements) {
            //拼接URL
            String product_url = "https://www.*****.co.za"
                    + ele.attr("href");
            next.add(product_url)   //添加到下面访问的URL
                    .type("ThirdLayer");    //第三层次
        }

    }

    /**
     * 针对每个产品
     * 解析得到产品的具体参数信息,即产品id
     * 产品描述和产品型号
     */
    public static void getParData(Page page, List<ParameterModel>
listPar) {
        String product_id = page
                .select("span[itemprop=productId]").text();
        String description = page
                .select("span[itemprop=description]").text();
        String model = page
                .select("span[itemprop=model]").text();
        ParameterModel pModel = new ParameterModel();
        pModel.setProduct_id(product_id);
        pModel.setDescription(description);
```

```
        pModel.setModel(model);;
        listPar.add(pModel);
    }
}
```

（2）com.db

com.db 中包含两个 Java 文件，即 MyDataSource 与 MYSQLControl。MyDataSource 类的写法与第 7 章该类的写法一致。为存储品牌对应的产品信息和产品详情信息，需要在数据库（crawler）中，创建相应的数据表 makro_product 和 makro_parameter，以下为建表语句。

```
CREATE TABLE `makro_product` (
  `id` varchar(100) NOT NULL,
  `name` varchar(200) DEFAULT NULL,
  `price` varchar(200) DEFAULT NULL,
  `brand` varchar(200) DEFAULT NULL,
  `category` varchar(255) DEFAULT NULL,
  PRIMARY KEY (`id`)
) ENGINE=InnoDB DEFAULT CHARSET=utf8;

CREATE TABLE `makro_parameter` (
  `product_id` varchar(100) DEFAULT NULL,
  `description` text,
  `model` varchar(100) DEFAULT NULL
) ENGINE=InnoDB DEFAULT CHARSET=utf8;
```

针对待存储的这两种数据，需要在 MYSQLControl 类中构建两种方法，即 executeInsertPro() 及 executeInsertPar() 方法，如程序 9-22 所示。其中，executeInsertPro() 方法负责存储品牌对应的产品信息，executeInsertPar() 方法负责存储产品详情信息。

程序 9-22

```
package com.db;

import java.sql.SQLException;
import java.util.List;
import javax.sql.DataSource;
import org.apache.commons.dbutils.QueryRunner;
import com.model.ParameterModel;
import com.model.ProductModel;
public class MYSQLControl {
```

```java
      //根据自己的数据库地址修改
    static DataSource ds = MyDataSource.getDataSource
("jdbc:mysql://127.0.0.1:3306/"
            + "crawler?useUnicode=true&characterEncoding=UTF8");
    static QueryRunner qr = new QueryRunner(ds);
    public static void executeInsertPro(List<ProductModel> data) {
        Object[][] params = new Object[data.size()][5];  //数据的维度
        for ( int i = 0; i < params.length; i++ ){
            params[i][0] = data.get(i).getId();
            params[i][1] = data.get(i).getName();
            params[i][2] = data.get(i).getPrice();
            params[i][3] = data.get(i).getBrand();
            params[i][4] = data.get(i).getCategory();
        }
        //使用batch方法批量插入
        try {
            qr.batch("insert into makro_product(id,name,price,"
                    + "brand,category) "
                    + "values (?,?,?,?,?)", params);
            System.out.println("执行数据库完毕!" + "成功插入数据:"+data.size() + "条");
        } catch (SQLException e) {
            e.printStackTrace();
        }
    }
    public static void executeInsertPar(List<ParameterModel> data) {
        Object[][] params = new Object[data.size()][3];  //数据的维度
        for ( int i = 0; i < params.length; i++ ){
            params[i][0] = data.get(i).getProduct_id();
            params[i][1] = data.get(i).getDescription();
            params[i][2] = data.get(i).getModel();
        }
        //使用batch方法批量插入
        try {
            qr.batch("insert into makro_parameter(product_id,"
                    + "description,model) values (?,?,?)", params);
            System.out.println("执行数据库完毕!" + "成功插入数据:"+data.size() + "条");
        } catch (SQLException e) {
            e.printStackTrace();
```

```
            }
        }
    }
```

（4）com.main

最后，编写爬虫类 MakroCrawler，如程序 9-23 所示。MakroCrawler 类继承了 BreadthCrawler 类。在构造方法 MakroCrawler()中，利用种子 URL 实例化一个 CrawlDatum 对象，并将该对象的类型设置为 FirstLayer，即第一层次；之后，利用 addSeed(CrawlDatum datum)方法将 BreadthCrawler 对象添加为待采集的节点。

在 visit()方法中，需要判断页面的类型是否为 FirstLayer，在判断结果为 true 的情况下，解析每个品牌页面中的 URL，并使用 addAndReturn(String url)方法将 URL 添加为待采集的节点，同时需要将这一层所有 URL 标记为 SecondLayer 类型。之后，判断页面的类型是否为 SecondLayer，判断结果为 true 的情况下，调用 Parse 类中 getProductData()方法，解析数据并将数据封装到集合 listPro 中，同时将这一层解析得到的所有 URL 添加为待采集节点（标记为 ThirdLayer）。最后，判断页面类型是否为 ThirdLayer，如果判断结果为 true，则调用 Parse 类中的 getParData()方法，解析数据并将数据封装到集合 listPar 中。

在主方法中，设置了 URL 请求时间间隔、超时时间、线程数目和执行的深度。在执行完 start()方法之后，需要将封装的产品信息数据和产品详情数据插入具体的数据表，因此，这里分别调用了 executeInsertPro()方法及 executeInsertPar()方法。值得注意的是，为了保证线程安全，我们并没有在 visit()方法中执行数据库操作。

程序 9-23

```java
package com.main;

import cn.edu.hfut.dmic.webcollector.model.CrawlDatum;
import cn.edu.hfut.dmic.webcollector.model.CrawlDatums;
import cn.edu.hfut.dmic.webcollector.model.Page;
import cn.edu.hfut.dmic.webcollector.plugin.berkeley.BreadthCrawler;
import java.util.ArrayList;
import java.util.List;
import org.jsoup.nodes.Element;
import org.jsoup.select.Elements;
import com.db.MYSQLControl;
import com.model.ParameterModel;
import com.model.ProductModel;
```

```java
import com.parse.Parse;
/**
 * WebCollector数据入库案例
 *
 *
 *
 */
public class MakroCrawler extends BreadthCrawler {
    public static List<ProductModel> listPro = new ArrayList<ProductModel>();
    public static List<ParameterModel> listPar = new ArrayList<ParameterModel>();
    public MakroCrawler(String crawlPath) {
        super(crawlPath, true);
        String url = "https://www.*****.co.za"
                + "/electronics-computers/cellphones/c/BG";
        CrawlDatum datum = new CrawlDatum(url)
                .type("firstLayer"); //第一层
        this.addSeed(datum);
    }
    public void visit(Page page, CrawlDatums next) {
        /*
         * 第一层次的页面
         * 获取所有品牌的URL
         * 并以此为第二层页面的入口
         **/
        if (page.matchType("firstLayer")) {
            //解析第一层页面
            Elements results = page.
                    select("div[data-js-media=true]").select("a");
            for (Element ele: results) {
                String url_next = "https://www.*****.co.za"
                        + ele.attr("href");
                //添加第二层次需要访问的URL
                next.addAndReturn(url_next).type("SecondLayer");
            }
        }
        /*
         * 第二层页面
         * 获取每个品牌对应的产品基本信息
```

```java
        **/
        if (page.matchType("SecondLayer")) {
            //解析第二层次页面,并添加第三层次的页面URL
            Parse.getProductData(page, listPro, next);
        }
        if (page.matchType("ThirdLayer")) {
            //解析第三层次页面
            Parse.getParData(page, listPar);
        }
    }

    public static void main(String[] args) throws Exception {
        MakroCrawler crawler = new MakroCrawler("crawl_makro");
        //       crawler.setResumable(true);
        //执行间隔,这里robots有限制,需要遵循robots协议
        crawler.getConf().setExecuteInterval(1000);
        crawler.getConf().setConnectTimeout(10000); //连接超时
        crawler.getConf().setReadTimeout(10000);//读取数据超时
        crawler.setThreads(70);   //访问的线程数目
        crawler.start(5);   //执行的深度
        /**
         * 注意:如果数据量较大,可以在visit()方法中,每一条数据执行一次操作
         * 以保证线程安全
         * 如果在visit()方法中直接进行集合插入
         * 操作(List or Vector线程并不安全),然后清空
         * 会导致数据插入不全
         *
         * 针对少量的数据,可以在这里进行一次插入操作
         */
        MYSQLControl.executeInsertPro(listPro);
        MYSQLControl.executeInsertPar(listPar);
    }
}
```

执行程序 9-23,发现数据表 makro_product 和 makro_parameter 成功插入了一系列数据,如图 9.20 和图 9.21 所示。

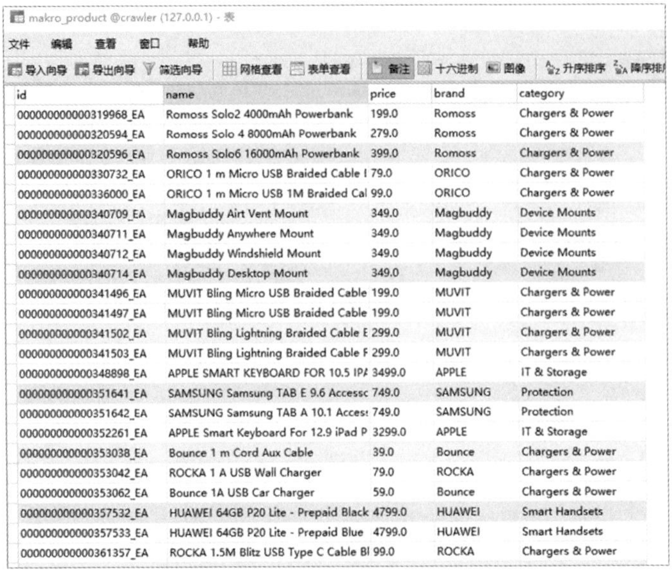

图 9.20 makro_product 表中存储的数据

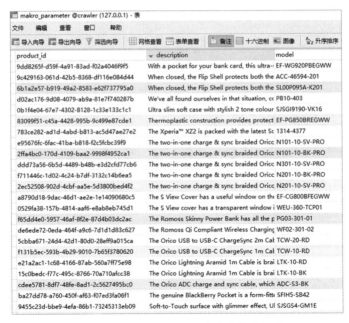

图 9.21 makro_parameter 表中存储的数据

9.3 WebMagic 的使用

9.3.1 WebMagic 简介

WebMagic 是一个非常优秀的 Java 开源网络爬虫框架，其功能覆盖了网络爬虫的整个生命周期，包括 URL 提取、网页内容下载、网页内容解析和数据存储。WebMagic 项目的源码可以在 GitHub 上进行下载。

WebMagic 具有很好的灵活性和可扩展性，支持多线程数据采集和 URL 去重等。

9.3.2 jar 包的下载

在 Maven 工程的 pom.xml 文件中可以配置 WebMagic 的最新 jar 包及其相应依赖 jar 包。

```
    <!-- https://************.com/artifact/us.codecraft/webmagic-core -->
    <dependency>
        <groupId>us.codecraft</groupId>
        <artifactId>webmagic-core</artifactId>
        <version>0.7.3</version>
    </dependency>
    <!-- https://************.com/artifact/us.codecraft/webmagic-extension -->
    <dependency>
        <groupId>us.codecraft</groupId>
        <artifactId>webmagic-extension</artifactId>
        <version>0.7.3</version>
    </dependency>
```

WebMagic 依赖的 jar 包有 HttpClient（网页请求开源库）、Jsoup/Xsoup（HTML 解析）、slf4j/log4j（配置日志）、Fastjson（JSON 数据解析）、jedis/commons-pool（Redis 数据库操作）等。

WebMagic 的核心模块包括 URL 维护/调度模块（Scheduler）、页面下载模块（Downloader）、URL 提取和页面分析模块（Processor）、数据解析模块（Selector）、数据存储模块（Pipeline）、代理操作模块（Proxy）。

9.3.3 入门案例（翻页数据采集）

下面，将以某大学新闻网为例，介绍 WebMagic 的使用情况。

使用 WebMagic 采集数据，只需要编写一个爬虫类，实现 PageProcessor 接口，如程序 9-24 所示。HFUTNewsProcessor 类实现了 PageProcessor 接口中的 getSite()和 process()方法。其中，getSite()方法用于配置网络爬虫，process()方法用于实现页面操作。

从程序 9-24 中，可以看到配置网络爬虫需要使用 Site 类中的相关方法，如调用 setRetryTimes()方法设置网页请求重试次数，调用 setSleepTime()方法设置页面请求之间的时间间隔。

在 process()方法中，首先判断请求的页面是新闻标题页还是新闻详情页（以"http://news.****.edu.cn/show."为前缀的 URL 为新闻详情页），如果是新闻标题页，则使用 Xpath 语法抽取所有新闻的 URL；同时，针对新闻标题页翻页的情况，如图 9.22 所示，使用 Xpath 路径表达式//*[@id='pages']//a[13]抽取页面中的"下一页"对应的 URL；最后，使用 Page 类中的 addTargetRequests()方法，将新闻详情页的 URL 和其他新闻标题页的 URL 添加到待请求 URL 队列中。如果页面是新闻详情页，则获取新闻 URL、新闻标题、新闻内容和新闻发表时间，同时，使用 Page 类中的 putField() 方法封装数据。在主方法中，需要使用 HFUTNewsProcessor 对象创建一个 Spider 对象，然后调用 addUrl()方法添加种子 URL，调用 thread()设置线程数目，调用 run()启动网络爬虫，调用 addPipeline(Pipeline pipeline)方法存储数据。为将结果以 JSON 的形式进行存储，这里使用 JsonFilePipeline。

程序 9-24

```java
package com.crawler.test;

import java.util.List;
import us.codecraft.webmagic.Page;
import us.codecraft.webmagic.Site;
import us.codecraft.webmagic.Spider;
import us.codecraft.webmagic.pipeline.JsonFilePipeline;
import us.codecraft.webmagic.processor.PageProcessor;
public class HFUTNewsProcessor implements PageProcessor {
    /**
     * 网络爬虫相关配置
     * 这里设置了重试次数，时间间隔
     */

    private Site site = Site.me().setRetryTimes(3).setSleepTime(1000);
    public Site getSite() {
        return site;
    }
```

```java
    /**
     * 针对每个URL对应的页面进行操作
     */
    public void process(Page page) {
        //如果不是以"http://news.****.edu.cn/show"为前缀的URL,进行下列操作
        if(!page.getUrl().regex("http://news.****.edu.cn/show.*").match()){
            //获取新闻URL
            List<String> urls = page.getHtml()
                    .xpath("//ul[@class='content list pushlist lh30']").links().all();
            //如果存在分页的情况,将每页URL添加到待采集的列表中
            List<String> url2 = page.getHtml()
                    .xpath("//*[@id='pages']//a[13]").links().all();
            urls.addAll(url2);
            page.addTargetRequests(urls);
        }else {   //使用Xpath语法解析数据
            String url = page.getUrl().toString();
            String title = page.getHtml().xpath("//*[@id='Article']/h2/text()").get();
            String content = page.getHtml().xpath("//*[@id='artibody']/allText()").get();
            String time = page.getHtml().xpath("//*[@id='Article']/h2/span/allText()")
                    .get();
            //存储结果
            page.putField("url",url);
            page.putField("title",title);
            page.putField("content",content);
            page.putField("time", time);
        }
    }
    public static void main(String[] args) {
        String url = "http://news.****.edu.cn/index.php?m=content&c=index&a=lists&catid=1";
        // 将该大学新闻网首页作为种子节点
        Spider.create(new HFUTNewsProcessor())
```

```
        .addUrl(url)
        //按照JSON的形式存储数据
        .addPipeline(new JsonFilePipeline("outputfile/"))
        .thread(5)   //开启5个线程抓取
        .run();      //启动爬虫
    }
}
```

图 9.22 该大学新闻网主页

运行程序 9-24，会发现在项目目录下的"outputfile"文件夹中自动创建了"news.****.edu.cn"文件夹，在该文件夹下产生了一系列 JSON 文件，其中每个 JSON 文件都对应一条新闻数据，如图 9.23 所示。

图 9.23 程序 9-24 采集的新闻数据

9.3.4 相关配置

在程序 9-24 中，使用 Site 类中的方法配置了网页请求的重试次数和页面请求之间的时间间隔。查看 Site 类的源码可以发现，它还可以配置网络爬虫的其他信息。

1. User-Agent

Site 类中的 setUserAgent()方法用于设置 User-Agent，如下为使用示例。

```
Site site = Site.me().setUserAgent("Your USER_AGENT");
```

2. 头信息

Site 类中的 addHeader ()方法用于设置请求头，如下为使用示例。

```
Site site = Site.me().addHeader("Accept-Encoding", "gzip, deflate")
        .addHeader("Accept-Language", "zh-CN,zh;q=0.9");
```

3. 超时时间

Site 类中的 setTimeOut ()方法用于设置超时时间，单位为毫秒，默认超时时间为 5000 毫秒，如下为使用示例。

```
Site site = Site.me().setTimeOut(3000);
```

4. 编码

Site 类中的 setCharset ()方法用于设置编码，如下为使用示例。

```
Site site = Site.me().setCharset("utf-8");
```

5. 循环重试次数

为防止某些网络原因导致 URL 请求失败，WebMagic 加入了循环重试机制。该机制会将请求失败的 URL 重新加入到待请求 URL 队列的尾部，继续请求，直到达到重试次数。为实现这个机制，需要调用 Site 类中的 setCycleRetryTimes ()方法，如下为该方法的使用示例。

```
Site site = Site.me().setCycleRetryTimes(3);
```

6. Cookie 添加

Site 类中的 addCookie ()方法可用于添加 Cookie 信息，如下为使用示例。

```
Site site = Site.me().addCookie("opxPID", "20190307");
```

9.3.5 数据存储方式

程序 9-24 使用了 JsonFilePipeline 类将解析的结果保存为 JSON 格式。另外，WebMagic，还提供了其他的 Pipeline 用于数据输出或存储，如 ConsolePipeline 和 FilePipeline。程序 9-25 和程序 9-26 分别给出了 ConsolePipeline 类和 FilePipeline 类中的 process()方法，从这两个程序中可以发现 ConsolePipeline 会将数据按行在控制台输出，FilePipeline 会将数据保存到 HTML 文件中。

程序 9-25

```java
public void process(ResultItems resultItems, Task task) {
    System.out.println("get page: " + resultItems.getRequest().getUrl());
    for (Map.Entry<String, Object> entry : resultItems.getAll().entrySet()) {
        System.out.println(entry.getKey() + ":\t" + entry.getValue());
    }
}
```

程序 9-26

```java
public void process(ResultItems resultItems, Task task) {
    String path = this.path + PATH_SEPERATOR + task.getUUID() + PATH_SEPERATOR;
    try {
        PrintWriter printWriter = new PrintWriter(new OutputStreamWriter(new
                FileOutputStream(getFile(path + DigestUtils.md5Hex(resultItems
                        .getRequest().getUrl()) + ".html")),
                "UTF-8"));
        printWriter.println("url:\t" + resultItems.getRequest().getUrl());
        for (Map.Entry<String, Object> entry : resultItems.getAll().entrySet()) {
            if (entry.getValue() instanceof Iterable) {
                Iterable value = (Iterable) entry.getValue();
                printWriter.println(entry.getKey() + ":");
                for (Object o : value) {
                    printWriter.println(o);
                }
```

```
            } else {
                printWriter.println(entry.getKey() + ":\t" + entry.
getValue());
            }
        }
        printWriter.close();
    } catch (IOException e) {
        logger.warn("write file error", e);
    }
}
```

ConsolePipeline 的使用方式如程序 9-27 所示，FilePipeline 的使用方式如程序 9-28 所示。

程序 9-27

```
Spider spider = Spider.create(new HFUTNewsProcessor())
        .addPipeline(new ConsolePipeline()); //控制台输出
spider.addUrl(url);
spider.thread(20).run();
```

程序 9-28

```
Spider spider = Spider.create(new HFUTNewsProcessor())
        //输出 HTML 文件
        .addPipeline(new FilePipeline("outputfile/"));
spider.addUrl(url);
spider.thread(20).run();
```

在实际应用中，用户可以通过定制 Pipeline 的方式实现其他数据存储需求，如将数据存储到 Excel 中等。程序 9-29 为定制的 Pipeline，其作用是将所采集的数据存储到一个文件中。为将 9.3.3 节中采集的新闻数据保存到指定的某个文件中，可以重写程序 9-24 中的主方法，如程序 9-30 所示，执行该主方法，可以看到"hfutnews.txt"文本文件成功插入了多条新闻数据，如图 9.24 所示。

程序 9-29

```
package com.crawler.test;

import java.io.BufferedWriter;
import java.util.Map;
import org.slf4j.Logger;
import org.slf4j.LoggerFactory;
```

```java
import us.codecraft.webmagic.ResultItems;
import us.codecraft.webmagic.Task;
import us.codecraft.webmagic.pipeline.Pipeline;
/**
 * 实现将采集的数据存储到一个文件中
 */
public class HFUTNewsOutPutFilePipeline implements Pipeline {
    private Logger logger = LoggerFactory.getLogger(getClass());
    public BufferedWriter w;
    public HFUTNewsOutPutFilePipeline(BufferedWriter writer) {
        this.w = writer;
    }
    public void process(ResultItems resultItems, Task task) {
        try {
            for (Map.Entry<String, Object> entry : resultItems.getAll().entrySet()) {
                w.append(entry.getKey() + ":\t" + entry.getValue() + "\n");
            }
            w.append("\n");
        } catch (Exception e) {
            logger.warn("write file error", e);
        }
    }
}
```

程序 9-30

```
        String url = "http://news.****.edu.cn/index.php?m=content&c=index&a=lists&catid=1";
        BufferedWriter writer = new BufferedWriter( new OutputStreamWriter
                ( new FileOutputStream(
                        new File("outputfile/hfutnews.txt")),
"utf-8"));
        Spider.create(new HFUTNewsProcessor())
        .addUrl(url)
        //数据保存到一个文件中
        .addPipeline(new HFUTNewsOutPutFilePipeline(writer))
        .thread(5)    //开启5个线程抓取
        .run();       //启动爬虫
        writer.close();
```

```
  2  url:      http://news.____.edu.cn/show-1-162437-1.html
  3  title:    我校学子在"第一届中国地球科学大数据挖掘与人工智能挑战赛"中喜获佳绩
  4  content:  近日，"第一届中国地球科学大数据挖掘与人工智能挑战赛"在中山大学落幕，来自40多所院校的
  5  time:     发布日期：2019-05-07  字号：大 中 小  【打印】
  6
  7  url:      http://news.____.edu.cn/show-1-162350-1.html
  8  title:    第十一届全国大学生广告艺术大赛（安徽赛区）启动仪式在我校举行
  9  content:  4月27日上午，第十一届全国大学生广告艺术大赛（安徽赛区）启动仪式暨名家名师校园创意巡
 10  time:     发布日期：2019-04-30  字号：大 中 小  【打印】
 11
 12  url:      http://news.____.edu.cn/show-1-162418-1.html
 13  title:    我校举办2019年毕业生春季第二场"双选会"
 14  content:  4月26日上午，我校2019年毕业生春季第二场"双选会"在屯溪路校区大学生活动中心成功举行。
 15  time:     发布日期：2019-05-05  字号：大 中 小  【打印】
 16
 17  url:      http://news.____.edu.cn/show-1-162315-1.html
 18  title:    我校在2019年中国高校创新创业学院联盟年会上荣获佳绩
 19  content:  4月22日～23日，教育部高等学校创新方法教学指导委员会主办的2019年中国高校创新创业学院
 20  time:     发布日期：2019-04-25  字号：大 中 小  【打印】
 21
 22  url:      http://news.____.edu.cn/show-1-162279-1.html
 23  title:    我校在中国仪器仪表学会成立40周年纪念大会暨学术年会上荣获多项表彰
 24  content:  日前，中国仪器仪表学会成立40周年纪念大会暨学术年会在北京隆重召开，来自仪器仪表、测量
 25  time:     发布日期：2019-04-24  字号：大 中 小  【打印】
 26
 27  url:      http://news.____.edu.cn/show-1-162244-1.html
 28  title:    我校学子在安徽省第五届大学生工程训练综合能力竞赛中再创佳绩
 29  content:  4月12-14日，安徽省第五届大学生工程训练综合能力竞赛暨第六届全国大学生工程训练综合能
 30  time:     发布日期：2019-04-23  字号：大 中 小  【打印】
```

图 9.24　程序 9-30 采集的新闻数据

9.3.6　数据采集入库

本节将使用 WebMagic 采集 ccm.net 网站中的软件的简介数据，如图 9.25 所示。首先，选取网站中的"DOWNLOAD"对应的 URL 作为第一个种子节点，如图 9.26 所示。另外，通过滚动条下拉页面，发现页面中只展示了部分软件。如果需要加载更多的软件，则必须单击网站底部的翻页按钮，如图 9.27 所示。基于浏览器抓包，发现在执行翻页操作时，服务器返回的数据以 HTML 格式存储，如图 9.28 所示，从该图中可以发现每款软件的 URL 均以"https://***.net/download/download-"为前缀；并且可以通过 h2[class='ccm_list_catch__item__title']进行定位；另外，该 HTML 数据对应的真实请求 URL 为 https://***.net/download/?page=2。其中，page 字段为页码编号，通过改变 page 字段的值，可以拼接出更多的 URL。将这些拼接的 URL 继续添加为种子节点，可以获取更多的软件简介数据。

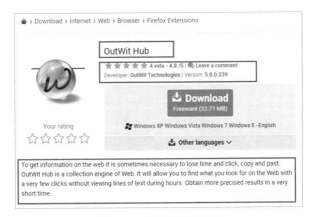

图 9.25　软件相关信息

图 9.26　DOWNLOAD 页面

图 9.27　查看更多软件

图 9.28　HTML 中对应的软件数据

1. 框架内容

图 9.29 所示为项目的结构，框架中的 package 有 com.（db、main、model、parse）。

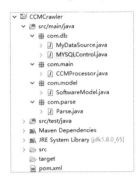

图 9.29　项目的结构

编写项目之前，需要使用 Maven 工程中的 pom.xml 文件配置整个项目需要的 jar 包。

```xml
        <!-- https://*************.com/artifact/us.codecraft/webmagic-core -->
        <dependency>
            <groupId>us.codecraft</groupId>
            <artifactId>webmagic-core</artifactId>
            <version>0.7.3</version>
        </dependency>
        <!-- https://*************.com/artifact/us.codecraft/webmagic-extension -->
        <dependency>
            <groupId>us.codecraft</groupId>
            <artifactId>webmagic-extension</artifactId>
            <version>0.7.3</version>
        </dependency>
        <!-- 数据库相关 -->
        <dependency>
            <groupId>mysql</groupId>
            <artifactId>mysql-connector-java</artifactId>
            <version>5.1.32</version>
        </dependency>
        <dependency>
            <groupId>commons-dbutils</groupId>
            <artifactId>commons-dbutils</artifactId>
            <version>1.7</version>
        </dependency>
        <dependency>
```

```
            <groupId>org.apache.commons</groupId>
            <artifactId>commons-dbcp2</artifactId>
            <version>2.5.0</version>
        </dependency>
```

2. 程序设计

（1）com.model

软件的简介信息包括：软件 id、软件名称、用户评论数量、用户评分、开发者、软件版本和软件描述（见图 9.25）。因此，在 com.model 下构建 Software 类用于封装需要存储的数据，如程序 9-31 所示。

程序 9-31

```java
package com.model;

public class SoftwareModel {
    private String id;                    //软件id
    private String name;                  //软件名称
    private String user_vote;             //用户评论数量
    private String voting;                //用户评分
    private String developer;             //开发者
    private String softwareVersion;       //软件版本
    private String description;           //软件描述
    public String getId() {
        return id;
    }
    public void setId(String id) {
        this.id = id;
    }
    public String getName() {
        return name;
    }
    public void setName(String name) {
        this.name = name;
    }
    public String getUser_vote() {
        return user_vote;
    }
    public void setUser_vote(String user_vote) {
        this.user_vote = user_vote;
    }
```

```java
    public String getVoting() {
        return voting;
    }
    public void setVoting(String voting) {
        this.voting = voting;
    }
    public String getDeveloper() {
        return developer;
    }
    public void setDeveloper(String developer) {
        this.developer = developer;
    }
    public String getSoftwareVersion() {
        return softwareVersion;
    }
    public void setSoftwareVersion(String softwareVersion) {
        this.softwareVersion = softwareVersion;
    }
    public String getDescription() {
        return description;
    }
    public void setDescription(String description) {
        this.description = description;
    }

}
```

（2）com.parse

程序 9-32 所示为解析类 Parse，其解析的页面如图 9.25 所示。图 9.25 中所示软件对应的 URL 为 https://***.net/download/download-2330-outwit-hub。在解析页面之前，需要使用浏览器抓包，分析 URL 对应的响应实体类型（如 HTML/JSON 等）及字段的定位。图 9.26 所示为网页抓包结果的部分截图。基于图 9.26 所示内容，可使用 CSS 选择器 span[itemprop= ratingCount]定位 ratingCount 字段（用户评分数目）、使用 span[itemprop= ratingValue]定位 ratingValue 字段（用户评分）、使用 dd[itemprop=author]定位 author 字段（软件开发者）；使用 Xpath 语法//dd[@itemprop='softwareVersion']定位软件版本。

在程序 9-32 中，getSoftData()方法中使用了 CSS 选择器和 Xpath 语法来抽取每款软件对应的数据。为将数据插入 MySQL 数据库，将解析的所有字段封装到实例化的 SoftwareModel 中。

程序 9-32

```java
package com.parse;
import com.model.SoftwareModel;
import us.codecraft.webmagic.Page;
public class Parse {
    /**
     * 针对每款软件的页面，解析获取软件的信息
     *
     * 这里使用CSS选择器及Xpath语法解析数据
     *
     */
    public static void getSoftData(Page page, SoftwareModel model) {
        //正则表达式取代所有非数字字符,获取软件id
        String id = page.getUrl()
                .replace("\\D", "").toString();
        String name = page.getHtml()
                .$("span[class=fn ftSize20]")
                .xpath("/allText()").get();
        String user_vote = page.getHtml()
                .$("span[itemprop=ratingCount]")
                .xpath("/allText()").get();
        String voting = page.getHtml()
                .$("span[itemprop='ratingValue']")
                .xpath("/allText()").get();
        String developer = page.getHtml()
                .$("dd[itemprop='author']")
                .xpath("/allText()").get();
        String sVersion = page.getHtml()
                .xpath("//dd[@itemprop='softwareVersion']/text()")
                .get();
        String description = page.getHtml()
                .xpath("//div[@itemprop='description']/allText()")
                .get();
        //封装数据
        model.setId(id);
        model.setName(name);
        model.setUser_vote(user_vote);;
        model.setVoting(voting);;
        model.setDeveloper(developer);
        model.setSoftwareVersion(sVersion);;
        model.setDescription(description);;
```

        }
    }

（3）com.db

com.db 中包含两个 Java 文件，即 MyDataSource 与 MYSQLControl。MyDataSource 类的写法与第 7 章该类的写法一致。为存储软件的相关信息，需要在数据库（crawler）中，创建数据表 ccmdata，以下为建表语句。

```sql
CREATE TABLE `ccmdata` (
  `id` int(10) NOT NULL,
  `name` varchar(100) DEFAULT NULL,
  `user_vote` varchar(10) DEFAULT NULL,
  `voting` varchar(10) DEFAULT NULL,
  `developer` varchar(100) DEFAULT NULL,
  `softwareVersion` varchar(10) DEFAULT NULL,
  `description` text,
  PRIMARY KEY (`id`)
) ENGINE=InnoDB DEFAULT CHARSET=utf8;
```

为执行数据插入操作，需要在 MYSQLControl 类中构建 executeInsertSoft() 方法，如程序 9-33 所示。

程序 9-33

```java
package com.db;

import java.sql.SQLException;
import javax.sql.DataSource;
import org.apache.commons.dbutils.QueryRunner;
import com.model.SoftwareModel;
public class MYSQLControl{
//根据本地数据库地址修改
    static DataSource ds = MyDataSource.getDataSource
("jdbc:mysql://127.0.0.1:3306/"
        + "crawler?useUnicode=true&characterEncoding=UTF8");
    static QueryRunner qr = new QueryRunner(ds);
    public static void executeInsertSoft(SoftwareModel data) {
        Object[][] params = new Object[1][7];   //数据的维度
        params[0][0] = data.getId();
        params[0][1] = data.getName();
        params[0][2] = data.getUser_vote();
        params[0][3] = data.getVoting();
        params[0][4] = data.getDeveloper();
```

```java
        params[0][5] = data.getSoftwareVersion();
        params[0][6] = data.getDescription();
        //使用batch方法批量插入
        try {
          qr.batch("insert into ccmdata(id,name,user_vote,"
              + "voting,developer,softwareVersion,"
              + "description) values (?,?,?,?,?,?,?)", params);
          System.out.println("成功插入一条数据!");
      } catch (SQLException e) {
          e.printStackTrace();
      }
    }
  }
}
```

（4）com.main

最后，编写爬虫类 CCMProcessor，如程序 9-34 所示。CCMProcessor 类实现了 PageProcessor 接口。在主方法中，首先使用 addUrl() 添加第一个种子节点；然后，利用循环的方式，拼接更多的 URL，并将这些 URL 添加为种子节点；最后，设置线程数目并启动爬虫。

在 process() 方法中，对 URL 进行如下判断:（a）如果 URL 以 "https://***.net/download/*" 为前缀，即 DOWNLOAD 页面和软件页面，则调用 getHtml() 方法获取 URL 对应的 HTML 内容，并利用 CSS 选择器和正则表达式提取页面中的所有软件的 URL，并将这些 URL 添加到待请求的 URL 队列中。（b）如果 URL 以 "https://***.net/download/download-.*" 为前缀，即软件简介页面，则调用 Parse 类（见程序 9-32）中的 getSoftData 方法()解析数据，最后，调用 MYSQLControl 类（见程序 9-33）中的 executeInsertSoft() 执行数据插入操作。

程序 9-34

```java
package com.main;

import java.io.IOException;
import java.util.List;
import com.db.MYSQLControl;
import com.model.SoftwareModel;
import com.parse.Parse;
import us.codecraft.webmagic.Page;
import us.codecraft.webmagic.Site;
import us.codecraft.webmagic.Spider;
import us.codecraft.webmagic.processor.PageProcessor;
```

```java
/**
 * HTML数据解析
 * 包含HTML和JSON的案例，可参考
 * https://******.com/soberqian/Java-Carwler-Technology/进行学习
 */
public class CCMProcessor implements PageProcessor {
    /**
     * 网络爬虫相关配置
     * 这里设置了重试次数、时间间隔
     */

    private Site site = Site.me()
            .setRetryTimes(3).setSleepTime(1000);
    public Site getSite() {
        return site;
    }
    /**
     * 针对每个URL对应进行判断操作
     */
    public void process(Page page) {
        //第一页为HTML
        if(page.getUrl().regex("https://***.net/download/*")
                .match()){
            /*
             * 获取软件对应的URL
             * 这里先通过CSS选择器进行定位
             * 之后利用正则表达式匹配每款软件对应的URL
             **/
            List<String> urls = page.getHtml()
                    .$("h2[class='ccm_list_catch__item__title']")
                    .links().regex("https://***.net/download/"
                            + "download-.*").all();
            page.addTargetRequests(urls);
        }
        //软件介绍页面
        if (page.getUrl().regex("https://***.net/download/"
                + "download-.*").match()) {
            SoftwareModel model = new SoftwareModel();
            //解析数据
            Parse.getSoftData(page, model);
            //存储数据，每一条数据存储一次
```

```java
            MYSQLControl.executeInsertSoft(model);
        }
    }

    public static void main(String[] args) throws IOException {
        //首页
        String url = "https://***.net/download/";
        Spider spider = Spider.create(new CCMProcessor());
        // 添加第一个种子节点
        spider.addUrl(url);
        //添加多页的URL，i从2开始循环
        for (int i = 2; i < 5; i++) {
            spider.addUrl("https://***.net/download/?page=" + i);
        }
        //开启5个线程抓取
        spider.thread(5);
        //启动爬虫
        spider.run();
    }
}
```

执行程序9-34，发现数据表 ccmdata 成功插入了一系列数据，如图9.30所示。

id	name	user_vote	voting	developer	softwareVersion	description
5310	Roxio Easy Media Cr	24	4.6	Roxio	9.0 (latest version)	Roxio Easy Media Creator is
6564	Opera Mini	5	4.6	Opera Software	7.0	Opera Mini helps users brow
9688	Internet Speed Boost	17	4.5	Download Upload	1.0.0.0 (latest versi	No updated version of this s
10327	xVideoServiceThief	21	4.5	XESC & TECHNOLOGY	2.5.2	xVideoServiceThief enables u
14629	YouTube Downloade	445	4.5	YouTube Downloader HD	2.9.9.58 (latest vers	YouTube Downloader HD is c
14957	Microsoft Fix It		3.8	Microsoft	2012	This software is no longer su
15575	TypingMaster Pro	192	4.5	TypingMaster Inc.	7.01	TypingMaster Pro is a softwa
16195	Samsung Kies	11	4.6	Samsung	3.1.13 (latest versic	Samsung Kies is a program t
22734	Free Timetabling Soft	11	4.5	Lalescu Liviu	5.12.2	Free Timetabling Software is
24084	Adobe Photoshop	810	4.6	Adobe	CC 2015	Adobe Photoshop is a photo
24107	Microsoft Save As PL	21	4.7	Microsoft	2	The Microsoft Save as PDF o
24159	Eclipse IDE for Java I	42	4.6	Eclipse Foundation	4.4.2	Eclipse is a software designe
24196	WhatsApp Messenge	10	4.4	WhatsApp Inc.	2.17.6	WhatsApp Messenger for iPl
24197	Adobe Premiere Pro	228	4.6	Adobe Systems	4.3.0.260	Adobe Premiere Pro is a feat
24219	Microsoft Visio Pro	72	4.5	Microsoft	2016	The above download link for
24321	WhatsApp Messenge	39	4.1	WhatsApp Inc.	2.16.89	WhatsApp Messenger for Bla
24511	Minecraft Pocket Edit	273	4.7	Mojang	0.8.1	Minecraft: Pocket Edition is a
24569	CCM Live Forum for i	10	4.5	CCM Benchmark	2.00	CCM Live Forum is a lightwei
24747	Minecraft	393	4.6	Notch Development AB	1.10	Minecraft is a video game wl
24749	VidMate YouTube Dc	100	4.7	Nemo Studio	2.39	VidMate is an application for
24771	Google Play Store (A	741	4.6	Google Inc.	6.7.13	Google Play Store is Google'
24777	Instagram for PC	320	4.6	Instagram	10.15.0	Instagram is a social media r
24838	Waptrick for Androic	1	5.0	JBStudio	1	waptrick.com is a web portal
24840	Facebook Lite for An	6	2.3	Facebook	12.0.0.5.140	Facebook Lite is, like its nam
24842	Facebook for Windo	(Null)	(Null)	Facebook	Varies	Facebook for Windows Phon

图 9.30　ccmdata 表存储的数据

9.3.7 图片的采集

本节将使用 WebMagic socwall（国外的一个图片分享网站）图库中的图片，如图 9.31 所示。采集的流程：拼接多个页面的 URL，并将这些 URL 添加为种子节点；获取每个种子节点对应的 HTML 页面，并解析页面中的所有的图片链接；使用 HttpClient 下载每张图片。

图 9.31　socwall 图库

在程序 9-35 中，ImageDownloaderUtil 为一个通用类，该类可用来下载图片、PDF 和压缩等文件。

程序 9-35

```
package com.crawler.image;
import org.apache.http.HttpResponse;
import org.apache.http.client.HttpClient;
import org.apache.http.client.methods.HttpGet;
import org.apache.http.impl.client.HttpClients;
import org.apache.http.util.EntityUtils;
import java.io.FileOutputStream;
import java.io.IOException;
import java.io.OutputStream;
import org.apache.log4j.Logger;
public class ImageDownloaderUtil {
  /**
   * 该方法使用 HttpClient下载图片、PDF、压缩文件等
   */
  private static Logger logger = Logger.getLogger
```

```
(ImageDownloaderUtil.class);
    public static synchronized void downLoadImage(String url,String fileName){
        //初始化HttpClient
        HttpClient httpClient = HttpClients.custom().build();
        HttpGet httpGet = new HttpGet(url);
        //获取结果
        HttpResponse httpResponse = null;
        try {
            httpResponse = httpClient.execute(httpGet);
        } catch (IOException e) {
            e.printStackTrace();
        }
        //非常简单的下载文件的方法
        try {
            OutputStream out = new FileOutputStream(fileName);
            httpResponse.getEntity().writeTo(out);
        } catch (Exception e) {
            logger.warn("write file error", e);
        }
        try {
            EntityUtils.consume(httpResponse.getEntity()); //消耗实体
        } catch (IOException e) {
            e.printStackTrace();
        }
    }
}
```

程序 9-36 所示为爬虫类，该类实现了 PageProcessor 接口。在 process()方法中，使用 CSS 选择器和 Xpath 语法抽取页面中所有图片的 URL，接着，循环每张图片的 URL，调用 ImageDownloaderUtil 类中的 downLoadImage()方法下载图片。

程序 9-36

```
package com.crawler.image;

import java.io.IOException;
import java.util.List;
import us.codecraft.webmagic.Page;
import us.codecraft.webmagic.Site;
import us.codecraft.webmagic.Spider;
```

```java
import us.codecraft.webmagic.processor.PageProcessor;
public class ImageProcessor implements PageProcessor {
    /**
     * 网络爬虫相关配置
     * 这里设置了重试次数、时间间隔
     */

    private Site site = Site.me().setTimeOut(3000);
    public Site getSite() {
        return site;
    }
    public void process(Page page) {
        //获取图片对应的URL
        List<String> url = page.getHtml().$("div[id=content]")
                .xpath("//div/a/img/@src").all();
        //下载每一张图片
        for (int i = 0; i < url.size(); i++) {
            String inputUrl = "https://www.*******.com" + url.get(i);
            ImageDownloaderUtil.downLoadImage(inputUrl,
                "image/" + inputUrl.split("/")[5] );
        }
    }
    public static void main(String[] args) throws IOException {
        long startTime = System.currentTimeMillis();
        Spider spider = Spider.create(new ImageProcessor());
        //通过URL拼接的方式采集多页
        for (int i = 1; i < 3; i++) {
            spider.addUrl("https://www.*******.com/wallpapers/page:" + i + "/");
        }
        spider.thread(5);   //开启5个线程抓取
        spider.run();  //启动爬虫
        long endTime = System.currentTimeMillis();
        System.ou.println("程序运行时间为: " + (endTime - startTime) + "ms");
    }
}
```

执行程序9-36，可以看到项目的"image"目录下成功下载了一系列图片，如图9.32所示。

图 9.32 程序 9-36 下载的图片

9.4 本章小结

本章主要介绍了 Crawler4j、WebCollector 和 WebMagic 三种网络爬虫框架的使用情况。但从适用性和可扩展性来看，笔者更倾向于利用 WebCollector 和 WebMagic 来开发数据采集项目。另外，这三种框架还有很多其他用法尚未介绍，如 JavaScript 加载数据的处理、分布式部署等，有兴趣的读者，可继续深入学习。